16 ⁵⁰/₄₀

Microbial Transformation

of

Steroids and Alkaloids

MICROBIAL TRANSFORMATION

OF

STEROIDS AND ALKALOIDS

by

HIROSHI IIZUKA, Ph. D.

Professor
Institute of Applied Microbiology
University of Tokyo

ATSUSHI NAITO, Ph. D.

Central Research Laboratories
Sankyo Co. Ltd.

UNIVERSITY OF TOKYO PRESS
Tokyo
UNIVERSITY PARK PRESS
State College, Pennsylvania

Published jointly by
UNIVERSITY OF TOKYO PRESS
Tokyo
and
UNIVERSITY PARK PRESS
State College, Pennsylvania
LIBRARY OF CONGRESS CATALOG CARD NUMBER 68—19102

Preface

In recent years, the most striking and significant development in the field of synthetic chemistry has been the application of biological systems to chemical reactions. Biological systems display a far greater specificity than the more conventional forms of organic chemical reactions, and of all the systems available, that which has the greatest immediate potential in organic synthesis is microbial transformation.

The first application of such a system was reported in 1937, in the work of Luigi Mamoli and Alberto Vercellone on the oxidative transformation of steroid compounds, using yeast and bacteria. The next few years saw little progress in the field, but in the mid 1950's there was a sudden interest in research on the problems of producing steroid hormones by microbial transformation. The method is significant both for basic research and for its industrial possibilities. At present, scientists are attempting the application of microbial transformation not only in the synthesis of steroid compounds, but more broadly for other organic compounds, such as alkaloids, etc., and a number of results have already been reported. Inevitably such efforts have led to an increased interest on the part of microbiologists in discovering and breeding of new strains, and these are constantly being searched for or developed in the laboratory. Thus it would seem that the new field of microbial transformation will develop more and more in the near future.

The present authors have been carrying out their studies on microbial transformation of steroids and alkaloids in the Institute of Applied Microbiology, the University of Tokyo, since 1956. This book presents a summary of the data which has been accumulated up to 1966.

The book itself is divided into ten chapters according to the type of substrate used in the microbial reactions. Within each chapter, the various transformation reactions are grouped by type. To simplify usage, representative substrates dealt with in each chapter are listed at the beginning, along with their chemical structures and structural names. For each specific reaction, the substrate is given on the left, and the reaction product on the right, as shown in Examples 1 and 2.

Below the substrate, the scientific name and strain name or number of the microorganism responsible for the particular reaction are given, followed by the percentage yield in parentheses. Below the reaction product, the name or names of the researchers and the reference for the

Example 1

Hydroxylation

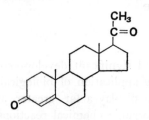 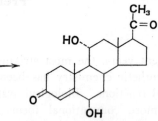

Progesterone 6β, 11α-Dihydroxyprogesterone

Aspergillus saitoi IAM R-1216   Iizuka, H., A. Naito and M. Hattori, J. Gen.
Appl. Microbiol. (Japan), **4**, 67 (1958)

Example 2

Dehydrogenation

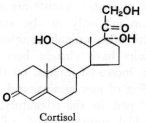 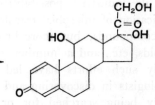

Cortisol 11β, 17α, 21-Trihydroxypregna-1, 4-
diene-3, 20-dione (Prednisolone)

Bacillus pulvifaciens IAM N-19-2   Iizuka, H., A. Naito and Y. Sato, J. Gen.
Appl. Microbiol (Japan), **7**, 118 (1961)

particular reaction are listed. In the section on steroid hormones, the hormones are arranged in order from C_{18}-steroids to C_{21}-steroids, within similar reaction patterns. Hydroxylation is shown according to the carbon number.

The microorganisms are identified by their scientific names, and the author's names of species are omitted, but the particular strains to which the strain name and number refer in the original reports are given. The index at the end of the book lists personal names, microorganisms, and compounds.

Because of the tremendous amount of work being done in the field at the present time, this book, by the time it leaves the press, will inevitably have a number of gaps in its coverage of recent research, making the subsequent publication of a revised edition imperative. In view of this, the authors would like to request anyone who has obtained any significant results on new microbial reactions relating to steroids, alkaloids or other substrates to send a copy of the paper or report to Hiroshi Iizuka, at the Institute of Applied Microbiology, the University of Tokyo, Bunkyo-ku, Tokyo, Japan.

The authors would also like to thank Dr. Hiroshi Okazaki for his kind advice with respect to collecting the materials for the present volume, and are deeply gratefull to all others who aided in the publication of this book.

Tokyo
September, 1967

Hiroshi Iizuka
and
Atsushi Naito

Contents

CONTENTS

CONTENTS

Microbial Transformation

of

Steroids and Alkaloids

I. MICROBIAL TRANSFORMATION OF STEROID HORMONES

Common names and systematic names of typical steroids

Common name	Systematic name
Androstenedione	Androst-4-ene-3,17-dione
Androsterone	3α-Hydroxy-5α-androstan-17-one
Corticosterone	11β, 21-Dihydroxypregn-4-ene-3,20-dione
Cortisol (Hydrocortisone)	11β, 17α, 21-Trihydroxypregn-4-ene-3,20-dione
Cortisone	17α, 21-Dihydroxypregn-4-ene-3,11,20-trione
11-Dehydrocorticosterone	21-Hydroxypregn-4-ene-3,11,20-trione
Dehydroepiandrosterone	3β-Hydroxyandrost-5-en-17-one
1-Dehydrotestololactone ($\varDelta^1$-Testololactone)	13α-Hydroxy-3-oxo-13,17-secoandrosta-1,4-dien-17-oic lactone
11-Deoxycorticosterone	21-Hydroxypregn-4-ene-3,20-dione
11-Deoxycortisol	17α, 21-Dihyroxypregn-4-ene-3,20-dione
	17α-Hydroxy-11-deoxycorticosterone
Estradiol	Estra-1,3,5(10)-triene-3, 17β-diol
Estrone	3-Hydroxyestra-1,3,5(10)-trien-17-one
Prednisolone	11β, 17α, 21-Trihydroxypregna-1,4-diene-3,20-dione
Prednisone	17α, 21-Dihydroxypregna-1,4-diene-3,11,20-trione
Pregnenolone	3β-Hydroxypregn-5-en-20-one
Progesterone	Pregn-4-ene-3,20-dione
Testololactone	13α-Hydroxy-3-oxo-13,17-secoandrost-4-en-17-oic lactone
Testosterone	17β-Hydroxyandrost-4-en-3-one

Structures of typical steroid hormones

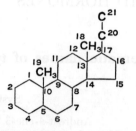

... α-configuration

— β-configuration

Basic skeleton of steroids

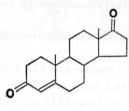

Androstenedione

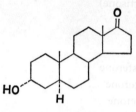

Androsterone

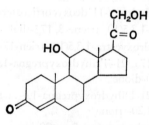

Corticosterone

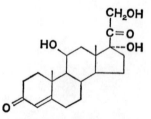

Cortisol (Hydrocortisone)

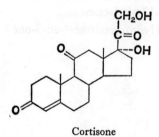

Cortisone

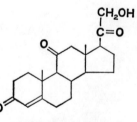

11-Dehydrocorticosterone

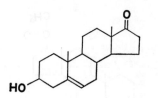

Dehydroepiandrosterone

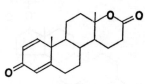

1-Dehydrotestololactone

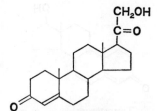

11-Deoxycorticosterone

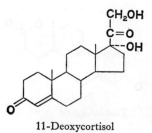

11-Deoxycortisol

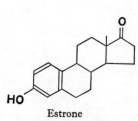

Estradiol   Estrone

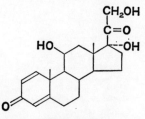

Prednisolone   Prednisone

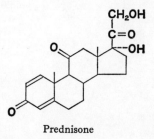

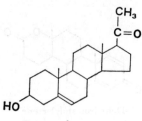

Pregnenolone

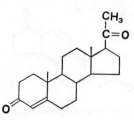

Progesterone

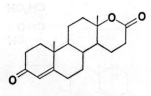

Testololactone

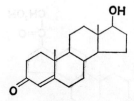

Testosterone

A. HYDROXYLATION

(a) 1-Hydroxylation

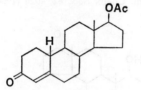  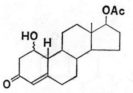

19-Nortestosterone 17-acetate   1 or 2-Hydroxy-19-nortestosterone
                                                  17-acetate

Corynebacterium simplex ATCC 6946   Kushinsky, S., J. Biol. Chem., **230**, 31 (1958)
(25%)

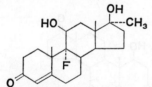 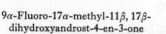 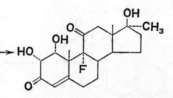

9α-Fluoro-17α-methyl-11β, 17β-   9α-Fluoro-17α-methyl-1α, 2α, 17β-
dihydroxyandrost-4-en-3-one   trihydroxyandrost-4-ene-3, 11-dione

Nocardia corallina ATCC 999 (18%)   Sax, K. J., C. E. Holmlund, L. I. Feldman, R. H.
                                                  Evans, Jr., R. H. Blank, A. J. Shay, J. S. Schultz
                                                  and M. Dann, Steroids, **5**, 345 (1965)

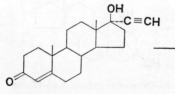  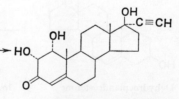

17α-Ethinyltestosterone   17α-Ethinyl-1α, 2α, 17β-
                                                  trihydroxyandrost-4-en-3-one

Nocardia corallina ATCC 999 (21%)   Sax, K. J., C. E. Holmlund, L. I. Feldman, R. H.
                                                  Evans, Jr., R. H. Blank, A. J. Shay, J. S. Schultz
                                                  and M. Dann, Steroids, **5**, 345 (1965)

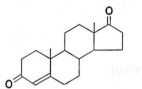 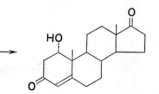

Androst-4-ene-3, 17-dione   1α-Hydroxyandrost-4-ene-3, 17-dione

Penicillium sp.

Dodson, R. M., A. H. Goldkamp and R. D. Muir, J. Am. Chem. Soc., **79**, 3921 (1957)

Penicillium sp. ATCC 12556

Dodson, R. M., A. H. Goldkamp and R. D. Muir, J. Am. Chem. Soc., **82**, 4026 (1960)

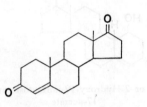 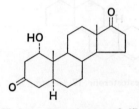

Androst-4-ene-3, 17-dione   1α-Hydroxy-5α-androstane-3, 17-dione

Penicillium sp. ATCC 12556

Dodson, R. M., A. H. Goldkamp and R. D. Muir, J. Am. Chem. Soc., **82**, 4026 (1960)

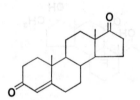 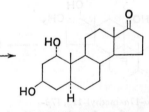

Androst-4-ene-3, 17-dione   1α, 3β-Dihydroxy-5α-androstan-17-one

Penicillium sp. ATCC 12556

Dodson, R. M., A. H. Goldkamp and R. D. Muir, J. Am. Chem. Soc., **82**, 4026 (1960)

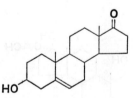 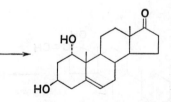

Dehydroepiandrosterone   1α, 3β-Dihydroxyandrost-5-en-17-one

Penicillium sp.

Dodson, R. M., A. H. Goldkamp and R. D. Muir, J. Am. Chem. Soc., **79**, 3921 (1957)

Penicillium sp. ATCC 12556

Dodson, R. M., A. H. Goldkamp and R. D. Muir, J. Am. Chem. Soc., **82**, 4026 (1960)

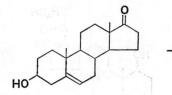 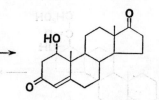

Dehydroepiandrosterone	1α-Hydroxyandrost-4-ene-3, 17-dione

Penicillium sp.

Dodson, R. M., A. H. Goldkamp and R. D. Muir, J. Am. Chem. Soc., **79**, 3921 (1957)

Penicillium sp. ATCC 12556

Dodson, R. M., A. H. Goldkamp and R. D. Muir, J. Am. Chem. Soc., **82**, 4026 (1960)

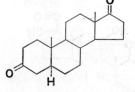 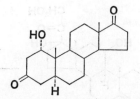

5α-Androstane-3, 17-dione	1α-Hydroxy-5α-androstane-3, 17-dione

Penicillium sp. ATCC 12556

Dodson, R. M., A. H. Goldkamp and R. D. Muir, J. Am. Chem. Soc., **82**, 4026 (1960)

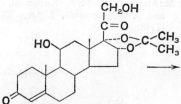 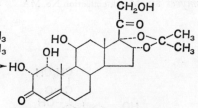

11β, 21-Dihydroxy-16α, 17α-isopropylidenedioxypregn-4-ene-3, 20-dione	1α, 2α, 11β, 21-Tetrahydroxy-16α, 17α-isopropylidenedioxypregn-4-ene-3, 20-dione

Nocardia corallina ATCC 999

Sax, K. J., C. E. Holmlund, L. I. Feldman, R. H. Evans, Jr., R. H. Blank, A. J. Shay, J. S. Schultz and M. Dann, Steroids, **5**, 345 (1965)

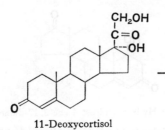

11-Deoxycortisol

1ξ, 17α, 21-Trihydroxypregn-
4-ene-3, 20-dione

Rhizoctonia ferrugena CBS

Greenspan, G., C. P. Schaffner, W. Charney, H. L.
Herzog and E. B. Hershberg, J. Am. Chem.
Soc., **79**, 3922 (1957)

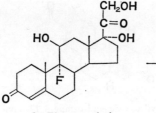

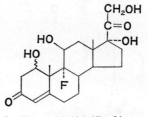

9α-Fluorocortisol

9α-Fluoro-1ξ, 11β, 17α, 21-
tetrahydroxypregn-
4-ene-3, 20-dione

Mortierella sp.

U. S. Pat. 2,962,423

Streptomyces sp. Merck collection No. MA
320

McAleer, W. J., M. A. Kozlowski, T. H. Stoudt
and J. M. Chemerda, J. Org. Chem., **23**, 508
(1958)

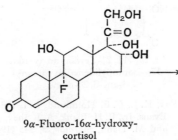

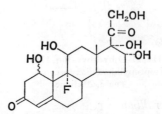

9α-Fluoro-16α-hydroxy-
cortisol

9α-Fluoro-1ξ, 11β, 16α, 17α, 21-
pentahydroxypregn-4-
ene-3, 20-dione

Mortierella sp.

U. S. Pat. 2,962,423

(b) 2-Hydroxylation

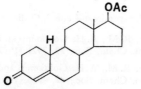

19-Nortestosterone 17-acetate

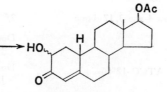

1 or 2-Hydroxy-19-nor-
testosterone 17-acetate

Corynebacterium simplex ATCC 6946
(25%)

Kushinsky, S., J. Biol. Chem., 230, 31 (1958)

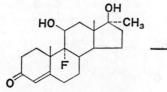

9α-Fluoro-17α-methyl-11β, 17β-
dihydroxyandrost-4-en-3-one

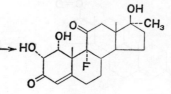

9α-Fluoro-17α-methyl-1α, 2α, 17β-
trihydroxyandrost-4-ene-
3, 11-dione

Nocardia corallina ATCC 999 (18%)

Sax, K. J., C. E. Holmlund, L. I. Feldman, R. H.
Evans Jr., R. H. Blank, A. J. Shay, J. S. Schultz
and M. Dann, Steroids, 5, 345 (1965)

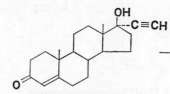

17α-Ethinyltestosterone

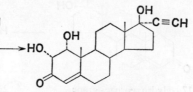

17α-Ethinyl-1α, 2α, 17β-
trihydroxyandrost-4-en-3-one

Nocardia corallina ATCC 999 (21%)

Sax, K. J., C. E. Holmlund, L. I. Feldman, R. H.
Evans, Jr., R. H. Blank, A. J. Shay, J. S. Schultz
and M. Dann, Steroids, 5, 345 (1965)

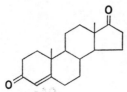

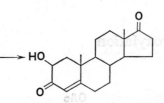

Androst-4-ene-3, 17-dione

2β-Hydroxyandrost-4-ene-3, 17-dione

Penicillium sp.

Dodson, R. M., A. H. Goldkamp and R. D. Muir, J. Am. Chem. Soc., **79**, 3921 (1957)

Penicillium sp. ATCC 12556

Dodson, R. M., A. H. Goldkamp and R. D. Muir, J. Am. Chem. Soc., **82**, 4026 (1960)

Streptomyces sp. DS-81-B

Herzog, H. L., M. J. Gentles, E. B. Hershberg, F. Carvajal, D. Sutter, W. Charney and C. P. Schaffner, J. Am. Chem. Soc., **79**, 3921 (1957)

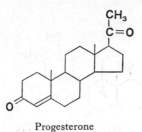

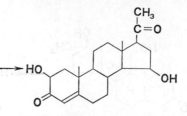

Progesterone

2β, 15β-Dihydroxyprogesterone

Sclerotinia libertiana

Tanabe, K., R. Takasaki, R. Hayashi and M. Shirasaka, Chem. Pharm. Bull. (Japan), **7**, 804 (1959)

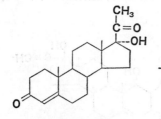

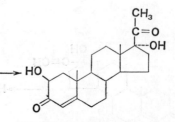

17α-Hydroxyprogesterone

2β, 17α-Dihydroxyprogesterone

Sclerotinia libertiana

Tanabe, K., R. Takasaki, R. Hayashi and M. Shirasaka, Chem. Pharm. Bull. (Japan), **7**, 804 (1959)

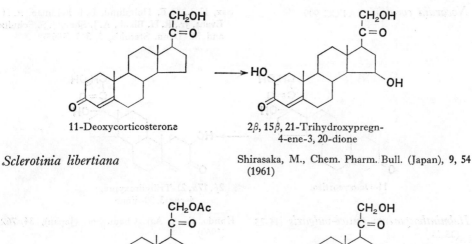

11-Deoxycorticosterone

2β, 15β, 21-Trihydroxypregn-
4-ene-3, 20-dione

Sclerotinia libertiana

Shirasaka, M., Chem. Pharm. Bull. (Japan), **9**, 54
(1961)

11-Deoxycorticosterone 21-
acetate

2β, 15β, 21-Trihydroxypregn-
4-ene-3, 20-dione

Sclerotinia sclerotiorum

Japan Pat. 311,627

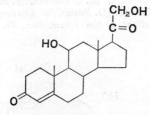 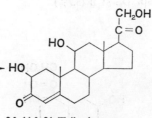

Corticosterone

2β, 11β, 21-Trihydroxypregn-
4-ene-3, 20-dione

Sclerotinia libertiana

Shirasaka, M., Chem. Pharm. Bull. (Japan), **9**, 54
(1961)

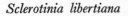

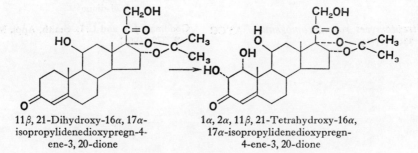

11β, 21-Dihydroxy-16α, 17α-
isopropylidenedioxypregn-4-
ene-3, 20-dione

1α, 2α, 11β, 21-Tetrahydroxy-16α,
17α-isopropylidenedioxypregn-
4-ene-3, 20-dione

Nocardia corallina ATCC 999

Sax, K. J., C. E. Holmlund, L. I. Feldman, R. H. Evans, Jr., R. H. Blank, A. J. Shay, J. S. Schultz and M. Dann, Steroids, **5**, 345 (1965)

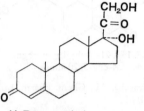

11-Deoxycortisol

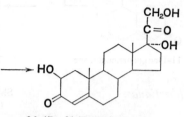

2β, 17α, 21-Trihydroxypregn-4-ene-3, 20-dione

Helminthosporium tritici-vulgaris H-25 (35%)

Kondo, E., J. Agr. Chem. Soc. (Japan), **34**, 762 (1960)

Rhizoctonia ferrugena CBS (3.2%)

Greenspan, G., C. P. Schaffner, W. Charney, H. L. Herzog and E. B. Hershberg, J. Am. Chem. Soc., **79**, 3922 (1957)

Rhizoctonia solani

U. S. Pat. 2,968,595

Sclerotinia libertiana

Tanabe, K., R. Takasaki, R. Hayashi and M. Shirasaka, Chem. Pharm. Bull. (Japan), **7**, 804 (1959)

Sclerotium oryzae

U. S. Pat. 2,968,595

Streptomyces sp. DS-81-B (5.7%)

Herzog, H.L., M. J. Gentles, E. B. Hershberg, F. Carvajal, D. Sutter, W. Charney and C. P. Schaffner, J. Am. Chem. Soc., **79**, 3921 (1957)

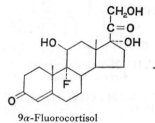

9α-Fluorocortisol

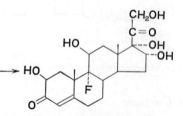

9α-Fluoro-2β, 11β, 16α, 17α, 21-pentahydroxypregn-4-ene-3, 20-dione

Streptomyces roseochromogenus ATCC 3347 (11%)

Goodman, J. J. and L. L. Smith, Appl. Microbiol., **9**, 372 (1961)

(c) 6-Hydroxylation

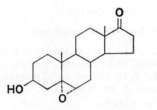

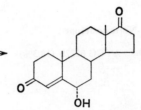

3β-Hydroxy-5α, 6α-oxido-
androstan-17-one

6α-Hydroxyandrost-4-ene-3, 17-dione

Nocardia restrictus No. 545

Lee, S. S. and C. J. Sih, Biochemistry, **3**, 1267
(1964)

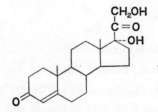

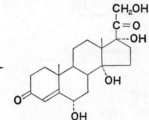

11-Deoxycortisol

6α, 14α, 17α, 21-Tetrahydroxypregn-
4-ene-3, 20-dione

Curvularia lunata

Japan pat. 311,865

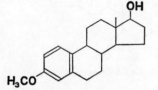

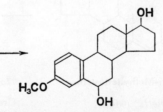

Estradiol 3-methylether

6β-Hydroxyestradiol 3-metylether

Fusarium moniliforme

Crabbé, P. and C. Casas-Campillo, J. Org. Chem.,
29, 2731 (1964)

Fusarium moniliforme ATCC 9851 (10%)

Fusarium moniliforme IH 4 (30%)

Casas-Campillo, C. and M. Bautista, Appl. Microbiol., **13**, 977 (1965)

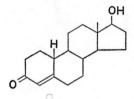 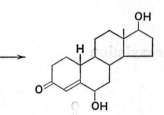

| 19-Nortestosterone | 6β-Hydroxy-19-nortestosterone |

Rhizopus nigricans ATCC 6227b (18.3%)   Pederson, R.L., J.A. Campbell, J.C. Babcock, S.H. Eppstein, H. C. Murray, A. Weintraub, R. C. Meeks, P. D. Meister, L. M. Reineke and D. H. Peterson, J. Am. Chem. Soc., **78**, 1512 (1956)

Rhizopus reflexus   U. S. Pat. 2,683,725

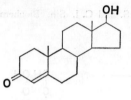 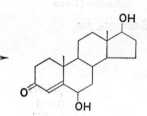

| Testosterone | 6β-Hydroxytestosterone |

Fusarium roseum   Rao, P. G., Indian J. Pharm., **25**, 131 (1963)

Rhizopus nigricans   Tamm, Ch., Angew. Chem., **74**, 225 (1962)

Rhizopus reflexus ATCC 1225   Eppstein, S. H., P. D. Meister, H. M. Leigh, D. H. Peterson, H. C. Murray, L. M. Reineke and A. Weintraub, J. Am. Chem. Soc., **76**, 3174 (1954)

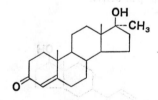 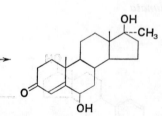

| 17α-Methyltestosterone | 17α-Methyl-6β-hydroxytestosterone |

Gibberella saubinetti (90%)   Urech, J., E. Vischer and A. Wettstein, Helv. Chim. Acta, **43**, 1077 (1960)

Rhizopus nigricans ATCC 6227b (4%)   Eppstein, S. H., P. D. Meister, H. M. Leigh, D. H. Peterson, H. C. Murray, L. M. Reineke and A. Weintraub, J. Am. Chem. Soc., **76**, 3174 (1954)

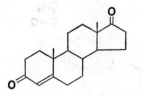 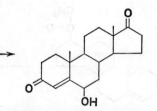

| Androst-4-ene-3, 17-dione | 6β-Hydroxyandrost-4-ene-3, 17-dione |

Aspergillus niger

Fried, J., R. W. Thoma, D. Perlman, J. E. Herz and A. Borman, Recent Progr. Hormone Res., **11**, 149 (1955)

Rhizopus arrhizus ATCC 11145

Eppstein, S. H., P. D. Meister, H. M. Leigh, D. H. Peterson, H. C, Murray, L. M. Reineke and A. Weintraub, J. Am. Chem. Soc., **76**, 3174 (1954)

Gibberella saubinetti (13%)

Urech, J., E. Vischer and A. Wettstein, Helv. Chim. Acta, **43**, 1077 (1960)

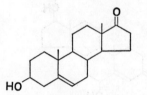 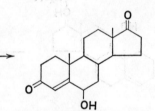

Dehydroepiandrosterone

6β-Hydroxyandrost-4-ene-3, 17-dione

Bacillus pulvifaciens IAM N-19-2

Iizuka, H., A. Naito, and Y. Sato, J. Gen. Appl. Microbiol. (Japan), 7, 118 (1961)

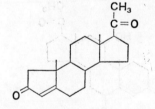 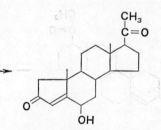

A-Norprogesterone

6β-Hydroxy-A-norprogesterone

Aspergillus nidulans

Japan Pat. 408,385

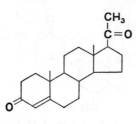

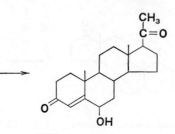

Progesterone

6β-Hydroxyprogesterone

Actinomyces sp.

Vondrová, O. and A. Capek, Folia Microbiol., **8**, 117 (1963)

Streptomyces aureofaciens

Fried, J., R. W. Thoma, D. Perlman, J. E. Herz and A. Borman, Recent Progr. Hormone Res., **11**, 149 (1955)

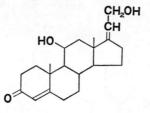

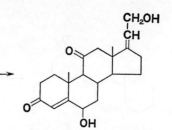

11β, 21-Dihydroxy-cis-pregna-4, 17(20)-dien-3-one

6β, 21-Dihydroxy-cis-pregna-4, 17 (20)-diene-3,11-dione

Rhizopus arrhizus

Hanze, A. R., O. K. Sebek and H. C. Murray, J. Org. Chem., **25**, 1968 (1960)

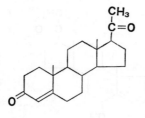

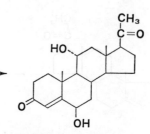

Progesterone

6β, 11α-Dihydroxyprogesterone

Aspergillus niger (45%)

Dulaney, E. L., W. J. McAleer, M. Koslowski, E. O. Stapley and J. Jaglom, Appl. Microbiol., **3**, 336 (1955)

Aspergillus niger strain Wisc. 72-2 (20%)

Fried, J., R. W. Thoma, J. R. Gerke, J. E. Herz, M. N. Donin and D. Perlman, J. Am. Chem. Soc., **74**, 3962 (1952)

Aspergillus ochraceus

Dulaney, E. L., Mycologia, **47**, 464 (1955)

Aspergillus ochraceus, conidia   Schleg, M. C. and S. G. Knight, Mycologia, **54**, 317 (1962)

Aspergillus saitoi IAM R-1216   Iizuka, H., A. Naito and M. Hattori, J. Gen. Appl. Microbiol. (Japan), **4**, 67 (1958)

Aspergillus terreus (41%)   Dulaney, E. L., W. J. McAleer, M. Koslowski, E. O. Stapley and J. Jaglom, Appl. Microbiol., **3**, 336 (1955)

Boletus luteus H-11
Dermoloma sp. F-27   Schuytema, E. C., M. P. Hargie, D. J. Siehr, I. Merits, J. R. Schenck, M. S. Smith and E. L. Varner, Appl. Microbiol., **11**, 256 (1963)

Gloeosporium kaki   Shirasaka, M. and M. Tsuruta, Chem. Pharm. Bull. (Japan), **9**, 159 (1961)

Hygrophorus conicus C-219
Leucopaxillus paradoxus F-55   Schuytema, E. C., M. P. Hargie, D. J. Siehr, I. Merits, J. R. Schenck, M. S, Smith and E. L. Varner, Appl. Microbiol., **11**, 256 (1963)

Rhizopus arrhizus ATCC 11145 (5~15%)   Peterson, D. H., H. C. Murray, S. H. Eppstein, L. M. Reineke, A. Weintraub, P. D. Meister and H. M. Leigh, J. Am. Chem. Soc., **74**, 5933 (1952)

Rhizopus cambodjae (30%)   Camerino, B., C. G. Alberti, A. Vercellone and F. Ammannati, Gazz. Chim. Ital., **84**, 301, (1954)

Rhizopus nigricans ATCC 6227b (2%)   Peterson, D. H., H. C. Murray, S. H. Eppstein, L. M. Reineke, A. Weintraub, P. D. Meister and H. M. Leigh, J. Am. Chem. Soc., **74**, 5933 (1952)

Sclerotium hydrophilum   Shirasaka, M. and M. Tsuruta, Chem. Pharm. Bull. (Japan), **9**, 196 (1961)

Streptomyces fradiae   Vondrová, O. et al., Folia Microbiol., **8**, 176 (1963)

Streptomyces sp.   Shirasaka, M. and M. Tsuruta, J. Ferm. Assoc. (Japan), **19**, 389 (1961)

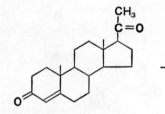 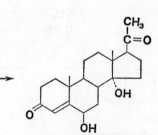

Progesterone 6β, 14α-Dihydroxyprogesterone

Achromobacter kashiwazakiensis IAM K-40-5   Tsuda, K., H. Iizuka, E. Ohki, Y. Sato, A. Naito and M. Hattori, J. Gen. Appl. Microbiol. (Japan), **5**, 7 (1959)

Bacillus cereus   Shirasaka, M., M. Ozaki and S. Sugawara, J. Gen. Appl. Microbiol. (Japan), **7**, 341 (1961)

Mucor corymbifer   Camerino, B., C. G. Alberti and A. Vercellone, Gazz. Chim. Ital., **83**, 684 (1953)

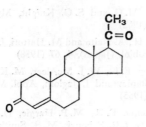 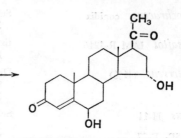

Progesterone   6β, 15α-Dihydroxyprogesterone

Fusarium lini   Gubler, A. and Ch. Tamm, Helv. Chim. Acta, **41**, 301 (1958)

Fusarium roseum   Rao, P. G., Indian J. Pharm., **25**, 131 (1963)

Gibberella saubinetti   Japan Pat. 313,770

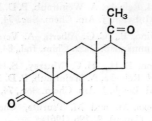 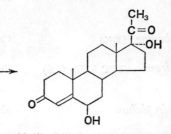

Progesterone   6β, 17α-Dihydroxyprogesterone

Naucoria confragosa C-172   Schuytema, E. C., M. P. Hargie, D. J. Siehr, I. Merits, J. R. Schenck, M. S. Smith and E. L. Varner, Appl. Microbiol., **11**, 256 (1963)

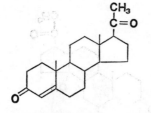 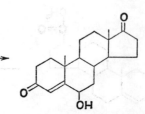

Progesterone   6β-Hydroxyandrost-4-ene-3, 17-dione

Gliocladium catenulatum ATCC 10523   Peterson, D. H., S. H. Eppstein, P. D. Meister, H. C. Murray, H. M. Leigh, A. Weintraub and L. M. Reineke, J. Am. Chem. Soc., **75**, 5768 (1953)

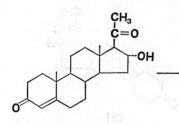 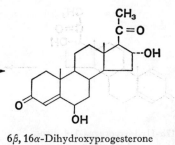

16α-Hydroxyprogesterone   6β, 16α-Dihydroxyprogesterone

Aspergillus nidulans

Fried, J., R. W. Thoma, D. Perlman, J. E. Herz and A. Borman, Recent Progr. Hormone Res., **11**, 149 (1955)

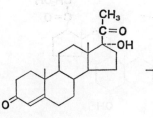 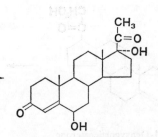

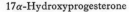

17α-Hydroxyprogesterone   6β, 17α-Dihydroxyprogesterone

Botrytis cinerea

Shirasaka, M., Chem. Pharm. Bull. (Japan), **9**, 152 (1961)

Fusarium lycopersici
Gibberella saubinetti

Shirasaka, M. and M. Tsuruta, Chem. Pharm. Bull. (Japan), **9**, 238 (1961)

Rhizopus arrhizus ATCC 11145 (45%)
Rhizopus nigricans ATCC 6227b (2%)

Meister, P. D., D. H. Peterson, H. C. Murray, G. B. Spero, S. H. Eppstein, A. Weintraub, L. M. Reineke and H. M. Leigh, J. Am. Chem. Soc., **75**, 416 (1953)

Sclerotium hydrophilum

Shirasaka, M. and M. Tsuruta, Chem. Pharm. Bull. (Japan), **9**, 196 (1961)

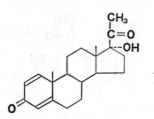
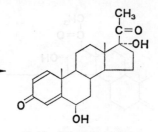

17α-Hydroxypregna-1, 4-diene-
3, 20-dione

6β, 17α-Dihydroxypregna-
1, 4-diene-3, 20-dione

Chaetomium funicola

Ger. Pat. 1,095,278

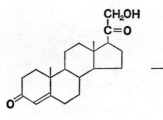
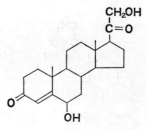

11-Deoxycorticosterone

6β, 21-Dihydroxypregn-
4-ene-3, 20-dione

Bacillus cereus

Shirasaka, M., M. Ozaki and S. Sugawara, J. Gen.
Appl. Microbiol. (Japan), **7**, 341 (1961)

Botrytis cinerea

Shirasaka, M., Chem. Pharm. Bull. (Japan), **9**, 152
(1961)

Lenzites abietina (4.8%)

Meystre, Ch., E. Vischer and A. Wettstein, Helv.
Chim. Acta, **38**, 381 (1955)

Rhizopus arrhizus ATCC 11145 (0.7%)

Eppstein, S. H., P. D. Meister, D. H. Peterson,
H. C. Murray, H. M. Leigh, D. A. Lyttle, L. M.
Reineke and A. Weintraub, J. Am. Chem. Soc.,
75, 408 (1953)

Sclerotinia sclerotiorum

Japan Pat. 311,629

Streptomyces fradiae

U. S. Pat. 2,649,401

Streptomyces sp.

Shirasaka, M. and M. Tsuruta, J. Ferm. Assoc.
(Japan), **19**, 389 (1961)

Trichothecium roseum

Meystre, Ch., E. Vischer and A. Wettstein, Helv.
Chim. Acta, **37**, 1548 (1954)

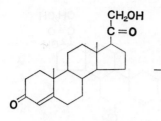
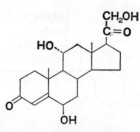

11-Deoxycorticosterone

6β, 11α, 21-Trihydroxypregn-
4-ene-3, 20-dione

Sclerotium hydrophilum

Shirasaka, M. and M. Tsuruta, Chem. Pharm. Bull.
(Japan), **9** 196 (1961)

Sclerotinia sclerotiorum

Japan Pat. 311,629

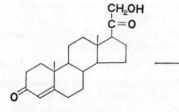
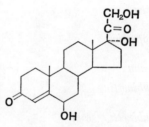

11-Deoxycorticosterone

6β, 17α, 21-Trihydroxypregn-
4-ene-3, 20-dione

Cephalothecium roseum ATCC 8685

Meister, P. D., L. M. Reineke, R. C. Meeks, H. C.
Murray, S. H. Eppstein, H. M. Leigh, A. Weintraub and D. H. Peterson, J. Am. Chem. Soc.,
76, 4050 (1954)

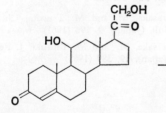
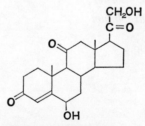

Corticosterone

6β, 21-Dihydroxypregn-
4-ene-3, 11, 20-trione

Sclerotium hydrophilum

Shirasaka, M. and M. Tsuruta, Chem. Pharm. Bull.
(Japan), **9**, 196 (1961)

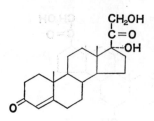 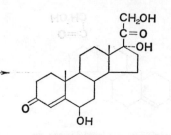

| 11-Deoxycortisol | 6β, 17α, 21-Trihydroxypregn-4-ene-3, 20-dione |

Achromobacter kashiwazakiensis IAM K-40-5

Tsuda, K., H. Iizuka, E. Ohki, Y. Sato, A. Naito and M. Hattori, J. Gen. Appl. Microbiol. (Japan), **5**, 7 (1959)

Bacillus cereus IAM B-204-1

Sugawara, S., M. Tsuruta, M. Shirasaka and M. Nakamura, Arch. Biochem. Biophys, **80**, 383 (1959)

Botrytis cinerea

Shirasaka, M., Chem. Pharm. Bull. (Japan), **9**, 152 (1961)

Fusarium lycopersici

Shirasaka, M. and M. Tsuruta, Chem. Pharm Bull. (Japan), **9**, 238 (1961)

Fusarium roseum

Rao P. G., Indian J. Pharm, **25**, 131 (1963)

Gibberella saubinetti

Shirasaka, M. and M. Tsuruta, Chem. Pharm. Bull. (Japan), **9**, 238 (1961)

Gibberella saubinetti (26%)

Urech J., E. Vischer and A. Wettstein, Helv. Chim. Acta, **43**, 1077 (1960)

Helicostylum piriforme

U. S. Pat. 2,602,769

Helminthosporium leersii

Kondo E., J. Agr. Chem. Soc. (Japan), **34**, 762 (1960)

Rhizopus arrhizus ATCC 11145 (20%)

Rhizopus nigricans ATCC 6227b

Peterson, D. H., S. H. Eppstein, P. D. Meister, B. J. Magerlein, H. C. Murray, H. M. Leigh, A. Weintraub and L. M. Reineke, J. Am. Chem. Soc., **75**, 412 (1953)

Sclerotinia sclerotiorum

Japan Pat. 311,629

Sclerotium hydrophilum

Shirasaka, M. and M. Tsuruta, Chem. Pharm. Bull. (Japan), **9**, 196 (1961)

Streptomyces sp.

Shirasaka, M. and M. Tsuruta, J. Ferm. Assoc. (Japan), **19**, 389 (1961)

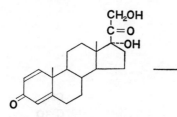
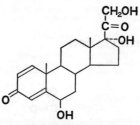

1-Dehydro-11-deoxycortisol

Chaetomium funicola

6β, 17α, 21-Trihydroxypregna-
1, 4-diene-3, 20-dione

Ger. Pat. 1,095,278

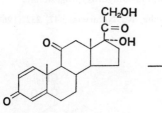
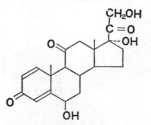

1-Dehydrocortisone (Prednisone)

Chaetomium funicola

6β, 17α, 21-Trihydroxypregna-
1, 4-diene-3, 11, 20-trione

Ger. Pat. 1,095,278

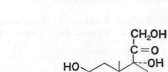
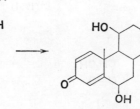
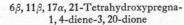

1-Dehydrocortisol
(Prednisolone)

Chaetomium funicola

6β, 11β, 17α, 21-Tetrahydroxypregna-
1, 4-diene-3, 20-dione

Ger. Pat. 1,095,278

(d) 7-Hydroxylation

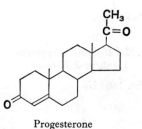

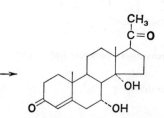

Progesterone

Curvularia lunata

7α, 14α-Dihydroxyprogesterone

Zetsche, K., Naturwiss., **47**, 232 (1960)

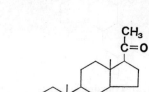

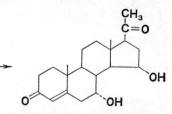

Progesterone

Helminthosporium sativum

7α, 15β-Dihydroxyprogesterone

Tsuda, K., T. Asai, Y. Sato, T. Tanaka and H. Hasegawa, Chem. Pharm. Bull. (Japan), **9**, 735 (1961)

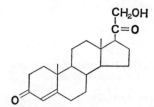

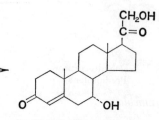

11-Deoxycorticosterone

7α, 21-Dihydroxypregn-
4-ene-3, 20-dione

Curvularia fallax (50~60%)
Curvularia pallescens (50~60%)
Peziza sp. ETH M-23 (50~60%)

Meystre, Ch., E. Vischer and A. Wettstein, Helv. Chim. Acta, **38**, 381 (1955)

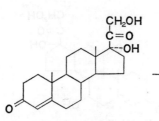

11-Deoxycortisol

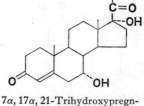

7α, 17α, 21-Trihydroxypregn-
4-ene-3, 20-dione

Cephalosporium sp.   U. S. Pat. 2,962,512

Diplodia natalensis   U. S. Pat. 2,960,436

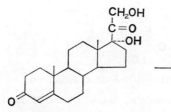

11-Deoxycortisol

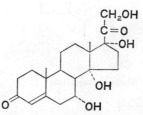

7α, 14α, 17α, 21-Tetrahydroxypregn-
4-ene-3, 20-dione

Curvularia lunata   Shull G. M., Trans. N. Y. Acad. Sci., **19**, 147 (1956)

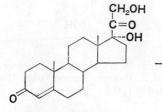

11-Deoxycortisol

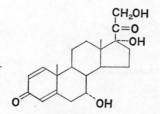

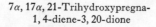

7α, 17α, 21-Trihydroxypregna-
1, 4-diene-3, 20-dione

Diplodia natalensis   U. S. Pat. 2,960,436

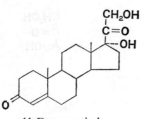

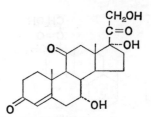

11-Deoxycortisol

7α, 17α, 21-Trihydroxypregn-
4-ene-3, 11, 20-trione

Diplodia natalensis

U. S. Pat. 2,960,436

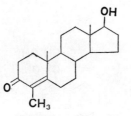

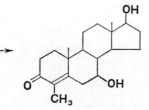

4-Methyltestosterone

4-Methyl-7β-hydroxytestosterone

Rhizopus nigricans

Tamm Ch., Angew. Chem., **74**, 225 (1962)

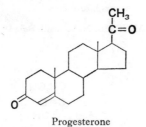

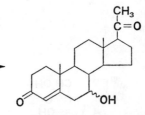

Progesterone

7ξ-Hydroxyprogesterone

Phycomyces blakesleeanus

Fried, J., R. W. Thoma, D. Perlman, J. E. Herz
and A. Borman, Recent Progr. Hormone Res.,
11, 149 (1955)

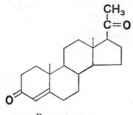

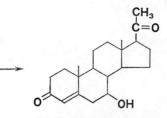

Progesterone

7β-Hydroxyprogesterone

Diplodia tubericola

Tsuda, K., T. Asai, Y. Sato, T. Tanaka and M.
Kato, J. Gen. Appl. Microbiol. (Japan), **5**, 1
(1959)

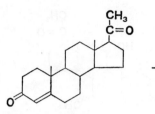

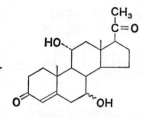

Progesterone	7ξ, 11α-Dihydroxyprogesterone

Absidia sp. (80%)

Rhizopus arrhizus (38%)

Ger. (East) Pat. 19,651

U. S. Pat. 2,602,769

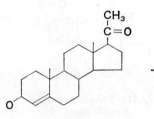

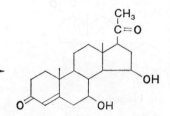

Progesterone	7β, 15β-Dihydroxyprogesterone

Helminthosporium sativum

Tsuda, K., T. Asai, Y. Sato, T. Tanaka and H. Hasegawa, Chem. Pharm. Bull. (Japan), **9**, 735 (1961)

Diplodia tubericola

Syncephalastrum racemosum

Tsuda, K., T. Asai, Y. Sato, T. Tanaka, T. Matsuhisa and H. Hasegawa, Chem. Pharm. Bull. (Japan), **8**, 626 (1960)

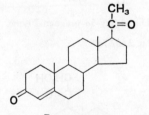

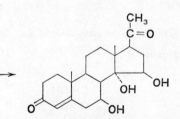

Progesterone	7β, 14α, 15β-Trihydroxypregn-4-ene-3, 20-dione

Syncephalastrum racemosum

Tsuda, K., T. Asai, Y. Sato, T. Tanaka, T. Matsuhisa and H. Hasegawa, Chem. Pharm. Bull. (Japan), **8**, 626 (1960)

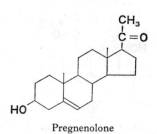

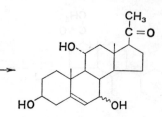

Pregnenolone

$3\beta, 7\xi, 11\alpha$-Trihydroxypregn-
5-en-20-one

Rhizopus arrhizus (17%)

Can. Pat. 506,689

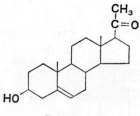

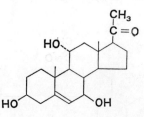

Pregnenolone

$3\beta, 7\beta, 11\alpha$-Trihydroxypregn-
5-en-20-one

Rhizopus arrhizus

Eppstein, S. H., P. D. Meister, H. C. Murray and
D. H. Peterson, Vitamines and Hormones, **14**,
359 (1956)

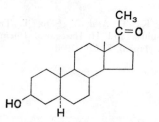

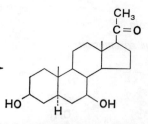

3β-Hydroxy-5α-pregnan-20-one

$3\beta, 7\beta$-Dihydroxy-5α-pregnan-20-one

Rhizopus arrhizus (5.2%)

U. S. Pat 2,602,769

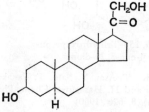

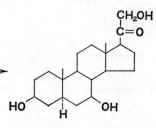

$3\beta, 21$-Dihydroxy-5α-pregnan-20-one

$3\beta, 7\beta, 21$-Trihydroxy-5α-pregnan-20-one

Rhizopus sp.

Kahnt, F. W., Ch. Meystre, R. Neher, E. Vischer
and A. Wettstein, Experientia, **8**, 422 (1952)

(e) 8-Hydroxylation

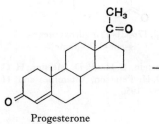 →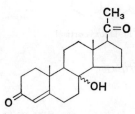

Progesterone

8 (or 9)-Hydroxyprogesterone

Streptomyces aureofaciens

Fried, J., R. W. Thoma, D. Perlman, J. E. Herz and A. Borman, Recent Progr. Hormone Res., **11**, 149 (1955)

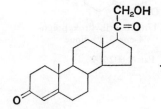 →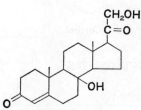

11-Deoxycorticosterone

8β, 21-Dihydroxypregn-
4-ene-3, 20-dione

Curvularia pallescens

Vischer, E., Ch. Meystre and A. Wettstein, Experientia, **11**, 465 (1955)

Helicostylum piriforme

Eppstein S. H., P. D. Meister, H. C. Murray and D. H. Peterson, Vitamines and Hormones, **14**, 359 (1956)

Mucor parasiticus
Neurospora crassa No. 74-A

Stone, D., M. Hayano, R. I. Dorfman, O. Hechter, C. R. Robinson and C. Djerassi, J. Am. Chem. Soc., **77**, 3926 (1955)

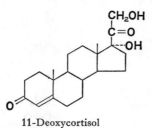

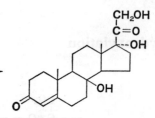

11-Deoxycortisol

8β, 17α, 21-Trihydroxypregn-
4-ene-3, 20-dione

Helicostylum piriforme

Eppstein, S. H., P. D. Meister, H. C. Murray and
D. H. Peterson, Vitamines and Hormones, **14**,
359 (1956)

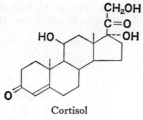

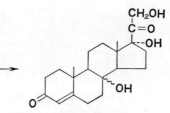

Cortisol

8ξ, 17α, 21-Trihydroxypregn-
4-ene-3, 20-dione

Helicostylum piriforme (8%)

U. S. Pat. 2,602,769

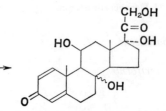

1-Dehydrocortisol
(Prednisolone)

8ξ, 11β, 17α, 21-Tetrahydroxypregna-
1, 4-diene-3, 20-dione

Helicostylum piriforme

Brit. Pat. 835,700

(f) 9-Hydroxylation

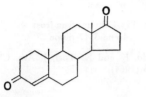

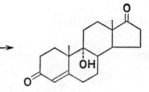

Androst-4-ene-3, 17-dione

9α-Hydroxyandrost-4-ene-3, 17-dione

Nocardia corallina

Nocardia restrictus

Brit. Pat. 862,701

Sih, C. J., Biochem. Biophys. Res. Comm., **7**, 87 (1962)

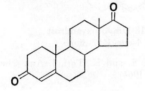

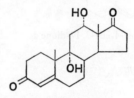

Androst-4-ene-3, 17-dione

9α, 12α-Dihydroxyandrost-4-ene-3, 17-dione

Cercospora melonis [*Corynespora melonis*]

Kondo, E. and K. Tori, J. Am. Chem. Soc., **86**, 736 (1964)

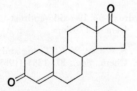

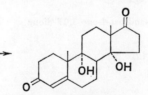

Androst-4-ene-3, 17-dione

9α, 14α-Dihydroxyandrost-4-ene-3, 17-dione

Cercospora melonis

Kondo, E. and K. Tori, J. Am. Chem. Soc., **86**. 736 (1964)

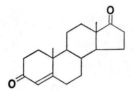

Androst-4-ene-3, 17-dione

Cercospora melonis

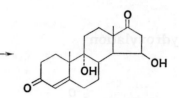

9α, 15β-Dihydroxyandrost-
4-ene-3, 17-dione

Kondo, E. and K. Tori, J. Am. Chem. Soc., **86**, 736 (1964)

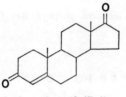

Androst-4-ene-3, 17-dione

Cercospora melonis

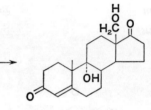

9α, 18-Dihydroxyandrost-
4-ene-3, 17-dione

Kondo, E. and K. Tori, J. Am. Chem. Soc., **86**, 736 (1964)

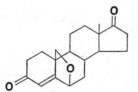

6β, 19-Oxido-androst-4-ene-3, 17-dione

Nocardia restrictus ATCC 14887

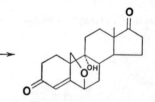

9α-Hydroxy-6β, 19-oxido-androst-
4-ene-3, 17-dione

Sih, C. J., S. S. Lee, Y. Y. Tsong and K. C. Wang, J. Am. Chem. Soc., **87**, 1385 (1965)

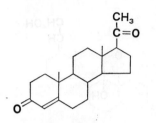

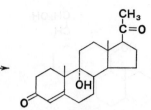

Progesterone

9α-Hydroxyprogesterone

Nocardia corallina
Nocardia restrictus

Brit. Pat. 862,701
U. S. Pat. 3,080,298

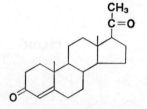

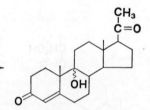

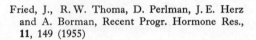

Progesterone

8 (or 9)-Hydroxyprogesterone

Streptomyces aureofaciens

Fried, J., R. W. Thoma, D. Perlman, J. E. Herz and A. Borman, Recent Progr. Hormone Res., **11**, 149 (1955)

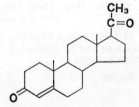

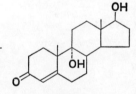

Progesterone

9α-Hydroxytestosterone

Nocardia corallina

Brit. Pat. 862,701

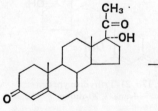

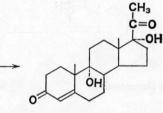

17α-Hydroxyprogesterone

9α, 17α-Dihydroxyprogesterone

Nocardia corallina

Brit. Pat. 862,701

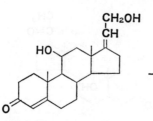 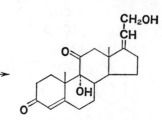

11β, 21-Dihydroxy-cis-pregna-
4, 17(20)-dien-3-one

9α, 21-Dihydroxy-cis-pregna-
4, 17(20)-diene-3, 11-dione

Helicostylum piriforme
Cunninghamella blakesleeana

Hanze, A. R., O. K. Sebek and H. C. Murray, J.
Org. Chem., **25**, 1968 (1960)

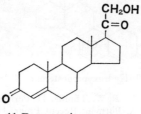 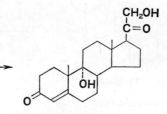

11-Deoxycorticosterone

9α, 21-Dihydroxypregn-
4-ene-3, 20-dione

Mucor parasiticus

Tamm, Ch., A. Gubler, G. Juhasz, E. Weiss-Berg
and W. Zürcher, Helv. Chim. Acta., **46**, 889
(1963)

Neurospora crassa No 74-A

Stone D., M. Hayano, R. I. Dorfman, O. Hechter,
C. R. Robinson and C. Djerassi, J. Am. Chem.
Soc., **77**, 3926 (1955)

Nocardia corallina

Brit. Pat. 862,701

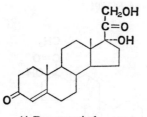 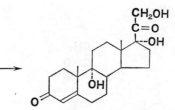

11-Deoxycortisol

9α, 17α, 21-Trihydroxypregn-
4-ene-3, 20-dione

Curvularia lunata (64%)
Nocardia corallina

Japan Pat. 417,538

Brit. Pat. 862,701

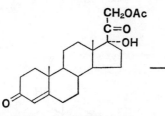 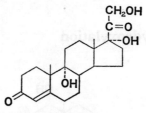

11-Deoxycortisol 21-acetate

9α, 17α, 21-Trihydroxypregn-
4-ene-3, 20-dione

Helminthosporium sigmoideum

Japan Pat. 417,540

(g) 10-Hydroxylation

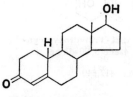

19-Nortestosterone

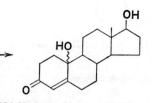

10ξ-Hydroxy-19-nortestosterone

Rhizopus nigricans   ATCC 6227b (1.2%)

Pederson, R. L., J. A. Campbell, J. C. Babcock, S. H. Eppstein, H. C. Murray, A. Weintraub, R. C. Meeks, P. D. Meister, L. M. Reineke and D. H. Peterson, J. Am. Chem. Soc., **78**, 1512 (1956)

(h) 11-Hydroxylation

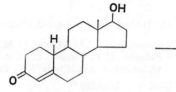 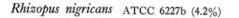 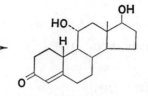

19-Nortestosterone	11α-Hydroxy-19-nortestosterone

Rhizopus nigricans ATCC 6227b (4.2%)

Pederson, R. L., J. A. Campbell, J. C. Babcock, S. H. Eppstein, H. C. Murray, A. Weintraub, R. C. Meeks, P. D. Meister, L. M. Reineke and D. H. Peterson, J. Am. Chem. Soc., **78**, 1512 (1956)

Rhizopus reflexus

U. S. Pat. 2,683,725

 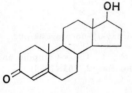 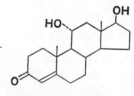

Testosterone	11α-Hydroxytestosterone

Aspergillus ochraceus NRRL 405, conidia

Vézina, C., S. N. Sehgal and K. Singh, Appl. Microbiol., **11**, 50 (1963)

Rhizopus nigricans

Tamm, Ch., Angew. Chem., **74**, 225 (1962)

Rhizopus reflexus ATCC 1225 (43%)

Eppstein, S. H., P. D. Meister, H. M. Leigh, D. H. Peterson, H. C. Murray, L. M. Reineke and A. Weintraub, J. Am. Chem. Soc., **76**, 3174 (1954)

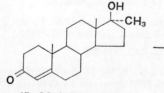 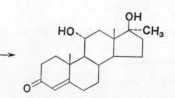

17α-Methyltestosterone	17α-Methyl-11α-hydroxytestosterone

Aspergillus ochraceus

Brit. Pat. 921,424

Rhizopus nigricans ATCC 6227b (47.5%)

Eppstein, S. H., P. D. Meister, H. M. Leigh, D. H. Peterson, H. C. Murray, L. M. Reineke and A. Weintraub, J. Am. Chem. Soc., **76**, 3174 (1954)

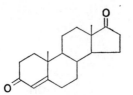

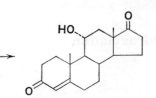

| Androst-4-ene-3, 17-dione | 11α-Hydroxyandrost-4-ene-3, 17-dione |

Rhizopus arrhizus ATCC 11145

Eppstein, S. H., P. D. Meister, H. M. Leigh, D. H. Peterson, H. C. Murray, L. M. Reineke and A. Weintraub, J. Am. Chem. Soc., **76**, 3174 (1954)

Rhizopus arrhizus (24%)
Rhizopus nigricans (24%)

U. S. Pat. 2,602,769

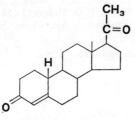

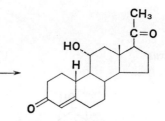

19-Norprogesterone · · · · · · · · · · 11α-Hydroxy-19-norprogesterone

Rhizopus nigricans ATCC 6227b

Bowers, A., C. Casas-Campillo and C. Djerassi, Tetrahedron, **2**, 165 (1958)

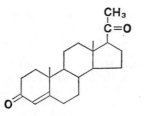

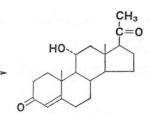

Progesterone · · · · · · · · · · · · · · · · 11α-Hydroxyprogesterone

Aspergillus awamori IAM K-0625 (46%)

Iizuka, H., A. Naito and M. Hattori, J. Gen. Appl. Microbiol. (Japan), **4**, 67 (1958)

Aspergillus itaconicus

U. S. Pat. 2,649,402

Aspergillus niger strain Wisc. 72-2 (35%)

Fried, J., R. W. Thoma, J. R. Gerke, J. E. Herz, M. N. Donin and D. Perlman, J. Am. Chem. Soc., **74**, 3962 (1952)

Aspergillus ochraceus (87%)

Dulaney, E. L., Mycologia, **47**, 464 (1955)

Aspergillus ochraceus (64%)

Brit. Pat. 912,274

Aspergillus ochraceus, conidia

Schleg, M. C. and S. G. Knight, Mycologia, **54**, 317 (1962)

Aspergillus ochraceus NRRL 405, conidia	Vézina, C., S. N. Sehgal and K. Singh, Appl. Microbiol., **11**, 50 (1963)
Aspergillus saitoi IAM R-1216 (70%)	Iizuka, H., A. Naito and M. Hattori, J. Gen. Appl. Microbiol. (Japan), **4**, 67 (1958)
Aspergillus sp.	Weisz, E., G. Wix and M. Bodánszky, Naturwiss., **43**, 39 (1956)
Aspergillus usamii U. V. mutant IAM 59-1 (18%)	Iizuka, H., A. Naito and M. Hattori, J. Gen. Appl. Microbiol. (Japan), **4**, 67 (1958)
Aspergillus usamii mut. *shirousamii* IAM B-407 (59%)	
Aspergillus wentii	U. S. Pat. 2,649,402
Bacillus cereus MB 717, NRRL B-1666	McAleer, W. J., T. A. Jacob, L. B. Turnbull, E. F. Schoenewaldt and T. H. Stoudt, Arch. Biochem. Biophys., **73**, 127 (1958)
Bacillus cereus var. *mycoides* MB 718	
Cnninghamella echinulata	U. S. Pat. 2,812,286
Dactylium dendroides (30%)	Dulaney, E. L., W. J. McAleer, H. R. Barkemeyer and C. Hlavac, Appl. Microbiol., **3**, 372 (1955)
Eurotium chevalieri (15%)	Brit. Pat. 740,858
Gloeosporium kaki	Shirasaka, M. and M. Tsuruta, Chem. Pharm. Bull. (Japan), **9**, 159 (1961)
Penicillium corylophilum	Dulaney, E. L., W. J. McAleer, M. Koslowski, E. O. Stapley and J. Jaglom, Appl. Microbiol., **3**, 336 (1955)
Penicillium lilacinum	
Penicillium tardum	
Pestalotia foedans (25%)	Can. Pat. 507,009
Pestalotia royenae	
Rhizopus arrhizus RH-176 (10%)	Peterson, D. H. and H. C. Murray, J. Am. Chem. Soc., **74**, 1871 (1952)
Rhizopus chinensis 10-10	Asai, T., K. Tsuda, K. Aida, E. Ohki, T. Tanaka, M. Hattori and H. Machida, J. Gen. Appl. Microbiol. (Japan), **4**, 63 (1958)
Rhizopus nigricans (80%)	U. S. Pat. 2,602,769
Rhizopus nigricans R-5-4	Asai, T., K. Tsuda, K. Aida, E. Ohki, T. Tanaka, M. Hattori and H. Machida, J. Gen. Appl. Microbiol. (Japan), **4**, 63 (1958)
Rhizopus sp. strain SY-152 (45%)	Mancera, O., A. Zaffaroni, B. A. Rubin, F. Sondheimer, G. Rosenkranz and C. Djerassi, J. Am. Chem. Soc., **74**, 3711 (1952)
Streptomyces fradiae	Vondrová, O. et al., Folia Microbiol (Prague), **8**, 176 (1963)
Streptomyces sp.	Shirasaka, M. and M. Tsuruta, J. Ferm. Assoc. (Japan), **19**, 389 (1961)

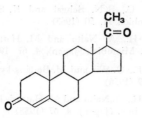 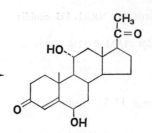

Progesterone	6β, 11α-Dihydroxyprogesterone
Aspergillus niger (45%)	Dulaney, E. L., W. J. McAleer, M. Koslowski, E. O. Stapley and J. Jaglom, Appl. Microbiol., **3**, 336 (1955)
Aspergillus niger strain Wisc. 72-2	Fried, J., R. W. Thoma, J. R. Gerke, J. E. Herz, M. N. Donin and D. Perlman, J. Am. Chem. Soc., **74**, 3962 (1952)
Aspergillus ochraceus	Dulaney, E. L., Mycologia, **47**, 464 (1955)
Aspergillus ochraceus, conidia	Schleg, M. C. and S. G. Knight, Mycologia, **54**, 317 (1962)
Aspergillus saitoi IAM R-1216	Iizuka, H., A. Naito and M. Hattori, J. Gen. Appl. Microbiol. (Japan), **4**, 67 (1958)
Aspergillus terreus (41%)	Dulaney, E. L., W. J. McAleer, M. Koslowski, E. O. Stapley and J. Jaglom, Appl. Microbiol., **3**, 336 (1955)
Aspergillus usamii U. V. mutant IAM 59-1	Iizuka, H., A. Naito and M. Hattori, J. Gen. Appl. Microbiol. (Japan), **4**, 67 (1958)
Boletus luteus H-11	Schuytema, E. C., M. P. Hargie, D. J. Siehr, I. Merits, J. R. Schenck, M. S. Smith and E. L. Varner, Appl. Microbiol., **11**, 256 (1963)
Dermoloma sp. F-27	
Gloeosporium kaki	Shirasaka, M. and M. Tsuruta, Chem. Pharm. Bull. (Japan), **9**, 159 (1961)
Hygrophorus conicus C-219	Schuytema, E. C., M. P. Hargie, D. J. Siehr, I. Merits, J. R. Schenck, M. S. Smith and E. L. Varner, Appl. Microbiol., **11**, 256 (1963)
Leucopaxillus paradoxus F-55	
Rhizopus arrhizus ATCC 11145	Peterson, D. H., H. C. Murray, S. H. Eppstein, L. M. Reineke, A. Weintraub, P. D. Meister and H. M. Leigh, J. Am. Chem. Soc., **74**, 5933 (1952)
Rhizopus cambodjae	Camerino, B., C. G. Alberti, A. Vercellone and F. Ammannati, Gazz. Chim. Ital., **84**, 301, (1954)
Rhizopus nigricans ATCC 6227b	Peterson, D. H., H. C. Murray, S. H. Eppstein, L. M. Reineke, A. Weintraub, P. D. Meister and H. M. Leigh, J. Am. Chem. Soc., **74**, 5933 (1952)
Sclerotium hydrophilum	Shirasaka, M. and M. Tsuruta, Chem. Pharm. Bull. (Japan), **9**, 196 (1961)

Streptomyces fradiae

Vondrová, O. et al., Folia Microbiol. (Prague), **8**, 176 (1963)

Streptomyces sp.

Shirasaka, M. and M. Tsuruta, J. Ferm. Assoc. (Japan), **19**, 389 (1961)

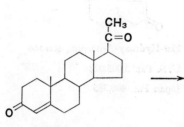 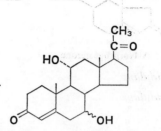

Progesterone   7ξ, 11α-Dihydroxyprogesterone

Absidia sp. (80%)   Ger. (East) Pat. 19,651

Rhizopus arrhizus (38%)   U. S. Pat. 2,602,769

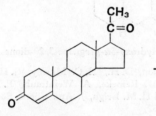 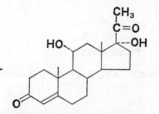

Progesterone   11α, 17α-Dihydroxyprogesterone

Cephalothecium roseum ATCC 8685

Meister, P. D., L. M. Reineke, R. C. Meeks, H. C. Murray, S. H. Eppstein, H. M. Leigh, A. Weintraub and D. H. Peterson, J. Am. Chem. Soc., **76**, 4050 (1954)

Dactylium dendroides (15.4%)

Dulaney, E. L., W. J. McAleer, H. R. Barkemeyer and C. Hlavac, Appl. Microbiol, **3**, 372 (1955)

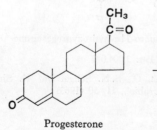 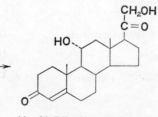

Progesterone   11α, 21-Dihydroxyprogesterone
                                     (11-Epicorticosterone)

Aspergillus sp.

Weisz, E., G. Wix and M. Bodánszky, Naturwiss., **43**, 39 (1956)

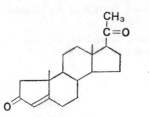

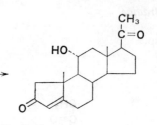

A-Norprogesterone

11α-Hydroxy-A-norprogesterone

Aspergillus nidulans

U. S. Pat. 3,005,028

Aspergillus nidulans

Japan Pat. 408,385

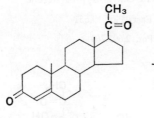

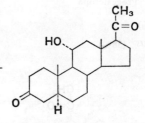

Progesterone

11α-Hydroxy-5α-pregnane-3, 20-dione

Rhizopus nigricans ATCC 6227b (0.5~
4.0%)

Peterson, D. H., H. C. Murray, S. H. Eppstein,
L. M. Reineke, A. Weintraub, P. D. Meister
and H. M. Leigh, J. Am. Chem. Soc., **74**, 5933
(1952)

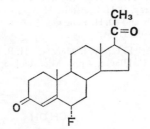

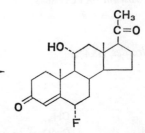

6α-Fluoroprogesterone

6α-Fluoro-11α-hydroxyprogesterone

Aspergillus nidulans

U. S. Pat. 3,004,047

Aspergillus ochraceus, conidia

Vézina, C., S. N. Sehgal and K. Singh, Appl.
Microbiol., **11**, 50 (1963)

— 44 —

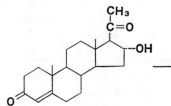 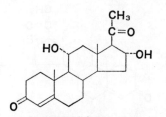

16α-Hydroxyprogesterone

11α, 16α-Dihydroxyprogesterone

Aspergillus nidulans (45%)

Fried, J., R. W. Thoma, D. Perlman, J. E. Herz and A. Borman, Recent Progr. Hormone. Res., **11**, 149 (1955)

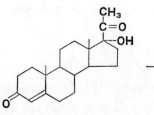 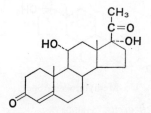

17α-Hydroxyprogesterone

11α, 17α-Dihydroxyprogesterone

Absidia regnieri

Shirasaka, M., Chem. Pharm. Bull. (Japan), **9**, 59 (1961)

Aspergillus niger strain Wisc. 72-2 (15%)

Fried, J., R. W. Thoma, J. R. Gerke, J. E. Herz M. N. Donin and D. Perlman, J. Am. Chem. Soc., **74**, 3962 (1952)

Aspergillus ochraceus

Dulaney, E. L., Mycologia, **47**, 464 (1955)

Aspergillus ochraceus, conidia

Vézina, C., S. N. Sehgal and K. Singh, Appl. Microbiol., **11**, 50 (1963)

Cunninghamella echinulata

U. S. Pat. 2,812,286

Dactylium dendroides (27%)

Dulaney, E. L., W. J. McAleer, H. R. Barkemeyer and C. Hlavac, Appl. Microbiol., **3**, 372 (1955)

Gloeosporium kaki

Shirasaka, M. and M. Tsuruta, Chem. Pharm. Bull. (Japan), **9**, 159 (1961)

Rhizopus arrhizus ATCC 11145 (2%)

Rhizopus nigricans ATCC 6227b (17%)

Meister, P. D., D. H. Peterson, H. C. Murray, G. B. Spero, S. H. Eppstein, A. Weintraub, L. M. Reineke and H. M. Leigh, J. Am. Chem. Soc., **75**, 416 (1953)

Sclerotinia libertiana

Shirasaka, M., Chem. Pharm. Bull. (Japan), **9**, 54 (1961)

Sclerotium hydrophilum

Shirasaka, M. and M. Tsuruta, Chem. Pharm. Bull. (Japan), 9, 196 (1961)

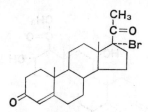

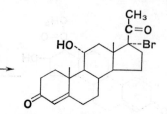

17α-Bromoprogesterone

17α-Bromo-11α-hydroxyprogesterone

Aspergillus ochraceus, conidia

Vézina, C., S. N. Sehgal and K. Singh, Appl. Microbiol., **11**, 50 (1963)

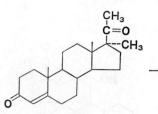

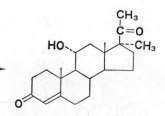

17α-Methylprogesterone

17α-Methyl-11α-hydroxyprogesterone

Aspergillus ochraceus, conidia

Vézina, C., S. N. Sehgal and K. Singh, Appl. Microbiol., **11**, 50 (1963)

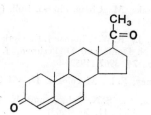

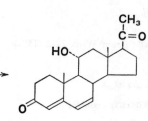

6-Dehydroprogesterone

11α-Hydroxypregna-4, 6-diene-3,20-dione

Rhizopus nigricans (50~60%)

Peterson, D. H., A. H. Nathan, P. D. Meister, S. H. Eppstein, H. C. Murray, A. Weintraub, L. M. Reineke and H. M. Leigh, J. Am. Chem. Soc., 75, 419 (1953)

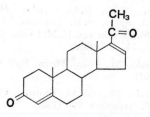

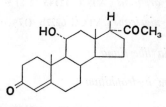

16-Dehydroprogesterone

11α-Hydroxy-17α-progesterone

Aspergillus niger

U. S. Pat. 2,649,402

Rhizopus nigricans

U. S. Pat. 2,602,769

Rhizopus nigricans ATCC 6227b

Meister, P. D., D. H. Peterson, H. C. Murray, S. H. Eppstein, L. M. Reineke, A. Weintraub and H. M. Leigh, J. Am. Chem. Soc., **75**, 55 (1953)

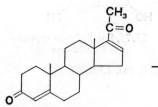

16-Dehydroprogesterone

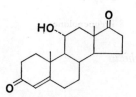

11α-Hydroxyandrost-4-ene-3, 17-dione

Aspergillus ochraceus, conidia

Vézina, C., S. N. Sehgal and K. Singh, Appl. Microbiol., **11**, 50 (1963)

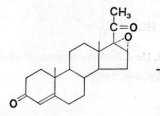

16α, 17α-Oxidoprogesterone

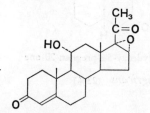

11α-Hydroxy-16α, 17α-oxidoprogesterone

Rhizopus nigricans (80%)

Ercoli, A., P. De Ruggieri and D. D. Morte, Gazz. Chim. Ital., **85**, 628 (1955)

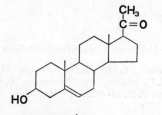

pregnenolone

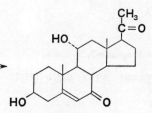

3β, 11α-Dihydroxypregn-5-ene-7, 20-dione

Rhizopus arrhizus (17.5%)

Can. Pat. 506,689

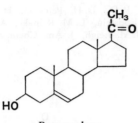

Pregnenolone

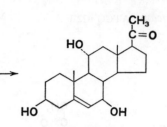

3β, 7β, 11α-Trihydroxypregn-
5-en-20-one

Rhizopus arrhizus

U. S. Pat. 2,702,809

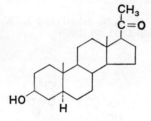

3β-Hydroxy-5α-pregnan-20-one

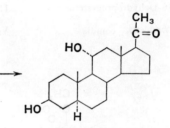

3β, 11α-Dihydroxy-5α-pregnan-20-one

Rhizopus nigricans (23%)

U. S. Pat. 2,602,769

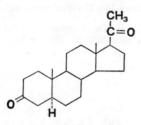

5α-Pregnane-3, 20-dione

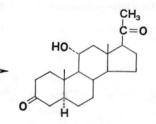

11α-Hydroxy-5α-pregnane-3, 20-dione

Rhizopus nigricans ATCC 6227b (25%)

Eppstein, S. H., D. H. Peterson, H. M. Leigh,
H. C. Murray, A. Weintraub, L. M. Reineke
and P. D. Meister, J. Am. Chem. Soc., **75**, 421
(1953)

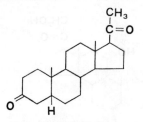

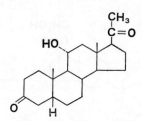

5β-Pregnane-3, 20-dione

11α-Hydroxy-5β-pregnane-3, 20-dione

Aspergillus niger

U. S. Pat. 2,649,402

Rhizopus nigricans ATCC 6227b (40%)

Eppstein, S. H., D. H. Peterson, H. M. Leigh,
H. C. Murray, A. Weintraub, L. M. Reineke and
P. D. Meister, J. Am. Chem. Soc., **75**, 421 (1953)

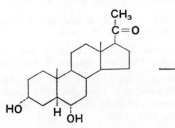

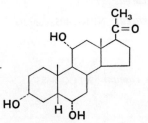

3α, 6α-Dihydroxy-5β-pregnan-20-one

3α, 6α, 11α-Trihydroxy-5β-pregnan-20-one

Calonectria decora

Ger. (East) Pat. 23,995

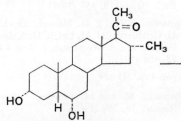

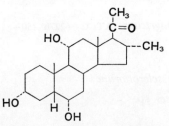

16α-Methyl-3α, 6α-dihydroxy-
5β-pregnan-20-one

16α-Methyl-3α, 6α, 11α-trihydroxy-
5β-pregnan-20-one

Calonectria decora

Ger. (East) Pat. 23,995

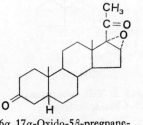

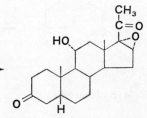

16α, 17α-Oxido-5β-pregnane-
3, 20-dione

11α-Hydroxy-16α, 17α-oxido-5β-pregnane-
3, 20-dione

Aspergillus ochraceus

U. S. Pat. 2,989,439

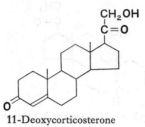

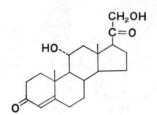

11-Deoxycorticosterone (Cortexone)	11α, 21-Dihydroxypregn-4-ene-3, 20-dione (11-Epicorticosterone)

Aspergillus niger strain Wisc. 72-2 (67%)
Fried, J., R. W. Thoma, J. R. Gerke, J. E. Herz, M. N. Donin and D. Perlman, J. Am. Chem. Soc., **74**, 3962 (1952)

Aspergillus ochraceus (75%)
Dulaney, E. L., Mycologia, **47**, 464 (1955)

Aspergillus ochraceus NRRL 405, conidia
Vézina, C., S. N. Sehgal and K. Singh, Appl. Microbiol., **11**, 50 (1963)

Bacillus cereus
Shirasaka, M., M. Ozaki and S. Sugawara, J. Gen. Appl. Microbiol. (Japan), **7**, 341 (1961)

Cephalothecium roseum ATCC 8685
Meister, P. D., L. M. Reineke, R. C. Meeks, H. C. Murray, S. H. Eppstein, H. M. Leigh, A. Weintraub and D. H. Peterson, J. Am. Chem. Soc., **76**, 4050 (1956)

Dactylium dendroides
Dulaney, E. L., W. J. McAleer, H. R. Barkemeyer and C. Hlavac, Appl. Microbiol., **3**, 372 (1955)

Glomerella lagenarium
Shirasaka, M. and M. Tsuruta, Chem. Pharm. Bull. (Japan), **9**, 159 (1961)

Rhizopus nigricans ATCC 6227b (50~60%)
Eppstein, S. H., P. D. Meister, D. H. Peterson, H. C. Murray, H. M. Leigh, D. A. Lyttle, L. M. Reineke and A. Weintraub, J. Am. Chem. Soc., **75**, 408 (1953)

Sclerotinia sclerotiorum
Japan Pat. 311,629

Streptomyces sp.
Shirasaka, M. and M. Tsuruta, J. Ferm. Assoc. (Japan), **19**, 389 (1961)

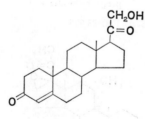

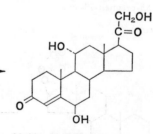

11-Deoxycorticosterone	6β, 11α, 21-Trihydroxypregn-4-ene-3, 20-dione

Sclerotinia sclerotiorum
Japan Pat. 311,629

Sclerotium hydrophilum
Shirasaka, M. and M. Tsuruta, Chem. Pharm. Bull. (Japan), **9**, 196 (1961)

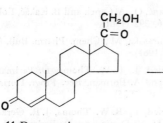

11-Deoxycorticosterone

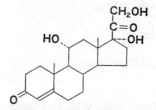

11α, 17α, 21-Trihydroxypregn-
4-ene-3, 20-dione

Cephalothecium roseum ATCC 8685

Meister, P. D., L. M. Reineke, R. C. Meeks, H. C.
Murray, S. H. Eppstein, H. M. Leigh, A. Weintraub and D. H. Peterson, J. Am. Chem. Soc.,
76, 4050 (1954)

Scopulariopsis brevicaulis

U. S. Pat. 2,970,085

Trichothecium roseum

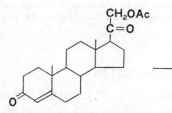

11-Deoxycorticosterone 21-acetate

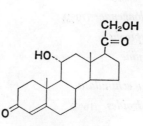

11α, 21-Dihydroxypregn-
4-ene-3, 20-dione

Aspergillus clavatus

U. S. Pat. 2,649,402

Aspergillus fischeri

Aspergillus nidulans (15.5%)

Aspergillus ustus

Rhizopus nigricans

U. S. Pat. 2,602,769

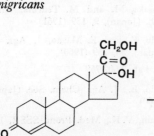

11-Deoxycortisol

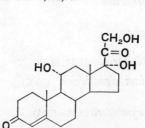

11α, 17α, 21-Trihydroxypregn-4-ene-
3, 20-dione (11-Epicortisol)

Absidia blakesleeana (18∼50%)

Eroshin, V. K., Med. Prom. SSSR, **16**, 23 (1962)

Absidia glauca

Schmidt-Thomé, J., Angew. Chem., **69**, 238 (1957)

Absidia orchidis (43.9%) Hanč, O., A. Čapek and B. Kakáč, Folia Microbiol., 6, 392 (1961)

Absidia regnieri Shirasaka, M., Chem. Pharm. Bull. (Japan), 9, 59 (1961)

Aspergillus nidulans (70%) Fried, J., R. W. Thoma, D. Perlman, J. E. Herz and A. Borman, Rec. Progr. Hormone Res., 11, 149 (1955)

Aspergillus niger strain Wisc. 72-2 (25%) Fried, J., R. W. Thoma, J. R. Gerke, J. E. Herz, M. N. Donin and D. Perlman, J. Am. Chem. Soc., 74, 3962 (1952)

Aspergillus ochraceus (50%) Dulaney, E. L., Mycologia, 47, 464 (1955)

Aspergillus ochraceus NRRL 405, conidia Vézina, C., S. N. Sehgal and K. Singh, Appl. Microbiol., 11, 50 (1963)

Bacillus cereus IAM B-204-1 Sugawara, S., M. Tsuruta, M. Shirasaka and M. Nakamura, Arch. Biochem. Biophys., 80, 383 (1959)

Beauveria sp. spore U. S. Pat. 3,013,945

Cercospora melongenae (70%) Kondo, E., K. Morihara, Y. Nozaki and E. Masuo, J. Agr. Chem. Soc. (Japan), 34, 844 (1960)

Cercospora scirpicola

Cercospora zinniae

Cunninghamella echinulata U. S. Pat. 2,812,286

Dactylium dendroides (16%) Dulaney, E. L., W. J. McAleer, H. R. Barkemeyer and C. Hlavac, Appl. Microbiol., 3, 372 (1955)

Delacroixia coronata (61%) Brit. Pat. 848,914

Didymella lycopersici ATCC 11847 (30%) Sehgal, S. N., K. Singh and C. Vézina, Steroids, 2, 93 (1963)

Didymella lycopersici, conidia Vézina, C., S. N. Sehgal and K. Singh, Appl. Microbiol., 11, 50 (1963)

Gloeosporium foliicolum Kondo, E. and E. Masuo, J. Agr. Chem. Soc. (Japan), 34, 759 (1960)

Glomerella cingulata

Glomerella lagenarium Shirasaka, M. and M. Tsuruta, Chem. Pharm. Bull. (Japan), 9, 159 (1961)

Glomerella mume Kondo, E. and E. Masuo, J. Agr. Chem. Soc. (Japan), 34, 759 (1960)

Helicostylum piriforme H-37, H-39 U. S. Pat. 2,602,769

Helminthosporium sigmoideum (70%) Kondo, E., J. Agr. Chem. Soc. (Japan), 34, 762 (1960)

Lichtheimia corymbifera (18~50%) Eroshin, V. K., Med. Prom. SSSR, 16, 23 (1962)

Lichtheimia ramosa (18~50%)

Mycocladus hyalinus (73~80%)

Penicillium expansum Dan. Pat. 91,515

Rhizopus nigricans (8%) U. S. Pat. 2,602,769

Sclerotinia sclerotiorum   Japan Pat. 311,629

Sclerotium hydrophilum   Shirasaka, M. and M. Tsuruta, Chem. Pharm.
                                 Bull. (Japan), **9**, 196 (1961)

Streptomyces sp.   Shirasaka, M. and M. Ozaki, J. Ferm. Assoc.
                                 (Japan), **19**, 389 (1961)

Tieghemella coerulea (23~24%) Eroshin, V. K., Med. Prom. SSSR, **16**, 23 (1962)

Tieghemella cylindrospora (73~80%)

Tieghemella hyalospora (73~80%)

Tieghemella orchidis (18~22%)

Tieghemella repens (73~80%)

Tieghemella spinosa (18~50%)

Tieghemella tiegemii (18~50%)

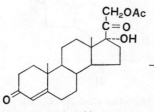

11-Deoxycortisol 21-acetate   11α, 17α, 21-Trihydroxypregn-4-ene-
                                   3, 20-dione (11-Epicortisol)

Dactylium dendroides   Dan. Pat. 94,041

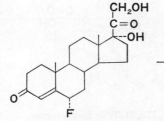 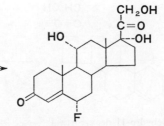

6α-Fluoro-11-deoxycortisol   6α-Fluoro-11α, 17α, 21-trihydroxypregn-
                                   4-ene-3, 20-dione

Aspergillus nidulans   U. S. Pat. 3,004,047

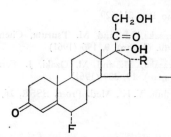

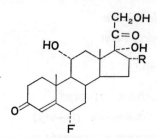

6α-Fluoro-16α-alkyl-11-deoxycortisol

6α-Fluoro-16α-alkyl-11α, 17α, 21-
trihydroxypregn-4-ene-3, 20-dione

Aspergillus ochraceus

U. S. Pat. 3,033,759

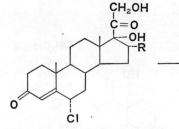

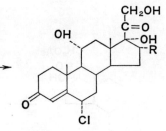

6α-Chloro-16α-alkyl-11-deoxycortisol

6α-Chloro-16α-alkyl-11α, 17α, 21-
trihydroxypregn-4-ene-3, 20-dione

Aspergillus ochraceus

U. S. Pat. 3,033,759

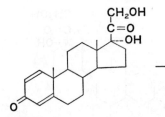

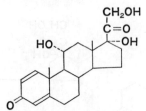

1-Dehydro-11-deoxycortisol

11α, 17α, 21-Trihydroxypregna-
1, 4-diene-3, 20-dione

Aspergillus ochraceus NRRL 405, conidia

Vézina, C., S. N. Sehgal and K. Singh, Appl.
Microbiol., **11**, 50, (1963)

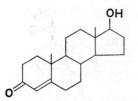

Testosterone

Aspergillus tamarii (25%)

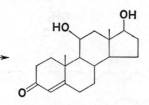

11β-Hydroxytestosterone

Brannon, D. R., J. Martin, A. C. Oehlschlager, N. N. Durham and L. H. Zalkow, J. Org. Chem., **30**, 760 (1965)

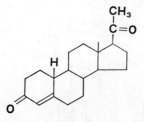

19-Norprogesterone

Curvularia lunata strain Syntex 192

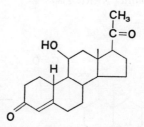

11β-Hydroxy-19-norprogesterone

Bowers, A., C. Casas-Campillo and C. Djerassi, Tetrahedron, **2**, 165 (1958)

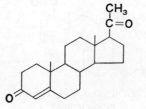

Progesterone

Aspergillus tamarii (14%)

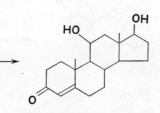

11β-Hydroxytestosterone

Brannon, D. R., J. Martin, A. C. Oehlschlager, N. N. Durham and L. H. Zalkow, J. Org. Chem., **30**, 760 (1965)

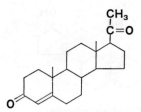

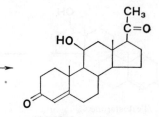

Progesterone

11β-Hydroxyprogesterone

Cunninghamella blakesleeana

Eppstein, S. H., P. D. Meister, H. C. Murray and D. H. Peterson, Vitamins and Hormones, **14,** 359 (1956)

Curvularia lunata NRRL 2380

Shull, G. M. and D. A. Kita, J. Am. Chem. Soc., **77,** 763 (1955)

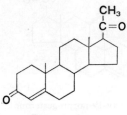

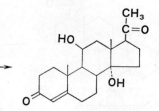

Progesterone

11β, 14α-Dihydroxyprogesterone

Curvularia lunata

Zetsche, K., Naturwiss., **47,** 232 (1960)

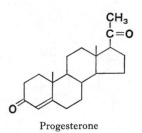

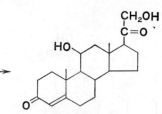

Progesterone

11β, 21-Dihydroxyprogesterone (Corticosterone)

Curvularia lunata

Rubin, B. A. et al., Bact. Proc., **56,** 33 (1956)

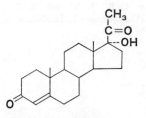

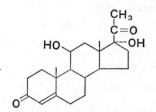

17α-Hydroxyprogesterone

11β, 17α-Dihydroxyprogesterone

Curvularia lunata NRRL 2380 (34%)

Shull, G. M. and D. A. Kita, J. Am. Chem. Soc., **77,** 763 (1955)

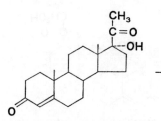

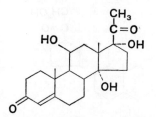

17α-Hydroxyprogesterone

Curvularia lunata NRRL 2380 (20%)

11β, 14α, 17α-Trihydroxyprogesterone

Shull, G. M. and D. A. Kita, J. Am. Chem. Soc., **77**, 763 (1955)

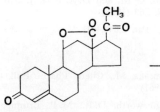

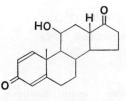

11β-Hydroxy-3, 20-dioxopregn-
4-en-18-oic acid-18, 11-lactone

Fusarium solani

11β-Hydroxy-18-norandrosta-
1, 4-diene-3, 17-dione

Urech, J., E. Vischer and A. Wettstein, Paper,
Meeting Swiss Chem. Soc., September, 1961

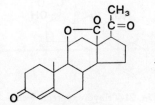

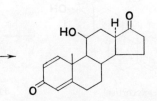

11β-Hydroxy-3, 20-dioxopregn-
4-en-18-oic acid-18, 11-lactone

Fusarium solani

11β-Hydroxy-18-nor-18-isoandrosta-
1, 4-diene-3, 17-dione

Urech, J., E. Vischer and A. Wettstein, Paper,
Meeting Swiss Chem. Soc., September, 1961

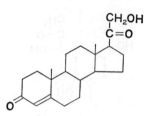 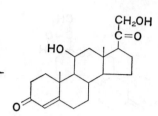

<table>
<tr><td>11-Deoxycorticosterone</td><td>11β, 21-Dihydroxypregn-4-ene-
3, 20-dione (Corticosterone)</td></tr>
</table>

Aspergillus fumigatus (2%)	U. S. Pat. 2,649,401
Cunninghamella blakesleeana H-334	Mann, K. M., F. R. Hanson and P. W. O'Connell, Federation Proc., **14**, 251 (1955)
Curvularia lunata NRRL 2380 (28%)	Shull, G. M. and D. A. Kita, J. Am. Chem. Soc., **77**, 763 (1955)
Penicillium chrysogenum (2%)	U. S. Pat. 2,649,401
Saccharomyces pastorianus (2%)	
Stachylidium bicolor	Shirasaka, M., Chem. Pharm. Bull. (Japan), **9**, 203 (1961)
Streptomyces fradiae strain Waksman 3535 (6%)	Colingsworth, D. R., M. P. Brunner and W. J. Haines, J. Am. Chem. Soc., **74**, 2381 (1952)
Streptomyces sp. (2%)	U. S. Pat. 2,649,401

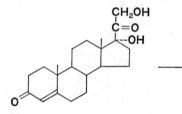 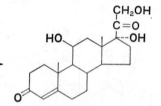

<table>
<tr><td>11-Deoxycortisol</td><td>11β, 17α, 21-Trihydroxypregn-4-ene-
3, 20-dione (Cortisol)</td></tr>
</table>

Absidia blakesleeana (0~38%)	Eroshin, V. K., Med. Prom. SSSR, **16**, 23 (1962)
Absidia glauca	Schmidt-Thomé, J., Angew. Chem., **69**, 238 (1957)
Absidia orchidis (48.9%)	Hanč, O., A. Čapek and B. Kakáč, Folia Microbiol., **6**, 392 (1961)
Botrytis cinerea	U. S. Pat. 2,789,940
Cephalothecium roseum	U. S. Pat. 2,765,258
Cercospora zinniae	Kondo, E., K. Morihara, Y. Nozaki and E. Masuo, J. Agr. Chem. Soc. (Japan), **34**, 844 (1960)
Colletotrichum phomoides	Brit. Pat. 749,414
Colletotrichum pisi	

Coniothyrium helliborine

Coniothyrium sp.

Eppstein, S. H., P. D. Meister, H. C. Murray and D. H. Peterson, Vitamines and Hormones, **14**, 359 (1956)

Corticium microsclerotia

Corticium praticola

Corticium sasakii IFO 5254

Hasegawa, T., T. Takahashi, M. Nishikawa and H. Hagiwara, Bull. Agr. Chem. Soc. (Japan), **21**, 390 (1957)

Corticium vagum

Japan Pat. 276,077

Cunninghamella blakesleeana H-334 (19%)

Mann, K. M., F. R. Hanson, P. W. O'Connell, H. V. Anderson, M. P. Brunner and J. N. Karnemaat, Appl. Microbiol., **3**, 14 (1955)

Curvularia falcata

U. S. Pat. 2,765,258

Curvularia lunata NRRL 2380 (40%)

Shull, G. M. and D. A. Kita, J. Am. Chem. Soc., **77**, 763 (1955)

Curvularia pallescens

U. S. Pat. 2,658,023

Dothichiza sp.

Shull, G. M., Trans. N. Y. Acad. Sci., **19**, 147 (1957)

Helminthosporium sativum

Japan Pat. 305,739

Helminthosporium sigmoideum H-40

Kondo, E., J. Agr. Chem. Soc. (Japan), **34**, 762 (1960)

Lichtheimia corymbifera (0~38%)

Lichtheimia ramosa (0~38%)

Mycocladus hyalinus (2~6%)

Eroshin, V. K., Med. Prom. SSSR, **16**, 23 (1962)

Pseudomonas fluorescens

Brit. Pat. 859,694

Pycnosporium sp.

Brit. Pat. 769,999

Rhodoseptoria sp.

Shull, G. M., Trans. N. Y. Acad. Sci., **19**, 147 (1957)

Stachylidium bicolor

Shirasaka, M., Chem. Pharm. Bull. (Japan), **9**, 203 (1961)

Streptomyces fradiae strain Waksman 3535 (6%)

Colingsworth, D. R., M. P. Brunner and W. J. Haines, J. Am. Chem. Soc., **74**, 2381 (1952)

Tieghemella coerulea (42~45%)

Eroshin, V.K., Med. Prom. SSSR, **16**, 23 (1962)

Tieghemella cylindrospora (2~6%)

Tieghemella hyalospora (2~6%)

Tieghemella orchidis (51~63%)

Tieghemella repens (2~6%)

Tieghemella spinosa (0~38%)

Tieghemella tiegemii (0~38%)

Trichothecium roseum

Brit. Pat. 749,414

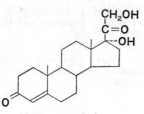

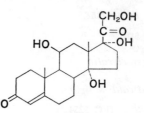

11-Deoxycortisol

11β, 14α, 17α, 21-Tetrahydroxypregn-
4-ene-3, 20-dione
(14α-Hydroxycortisol)

Curvularia lunata NRRL 2380

Agnello, E. J., B. L. Bloom and G. D. Laubach, J.,
Am. Chem. Soc., **77**, 4684 (1955)

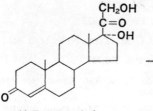

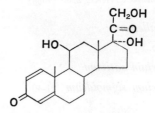

11-Deoxycortisol

11β, 17α, 21-Trihydroxypregna-1, 4-
diene-3, 20-dione (Prednisolone)

Absidia orchidis

Hung. Pat. 150,009

Mixed culture of *Corticium sasakii* and
Pseudomonas boreopolis

Japan Pat. 303,584

The actions of *Helminthosporium sativum*
and *Bacillus pulvifaciens* IAM N-19-2
in one and the same fermentation vessel in
sequence

U. S. Pat. 2,993,839

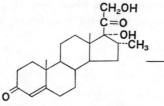

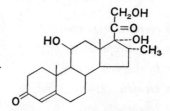

16α-Methyl-11-deoxycortisol

16α-Methyl-11β, 17α, 21-trihydroxy-
pregn-4-ene-3, 20-dione

Curvularia lunata

Ger. Pat. 1,147,226

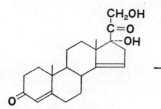

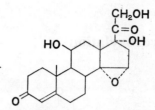

14-Dehydro-11-deoxycortisol

Curvularia lunata

$11\beta, 17\alpha, 21$-Trihydroxy-$14\alpha, 15\alpha$-oxidopregn-4-ene-3, 20-dione

Shull, G. M., Trans. N. Y. Acad. Sci., **19**, 147 (1956)

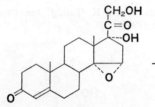

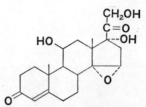

$14\alpha, 15\alpha$-Oxido-11-deoxycortisol

Curvularia lunata

$11\beta, 17\alpha, 21$-Trihydroxy-$14\alpha, 15\alpha$-oxidopregn-4-ene-3, 20-dione

Shull, G. M., Trans. N. Y. Acad. Sci., **19**, 147 (1956)

(i) 12-Hydroxylation

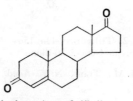

Androst-4-ene-3, 17-dione

Cercospora melonis

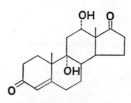

9α, 12α-Dihydroxyandrost-
4-ene-3, 17-dione

Kondo, E. and K. Tori, J. Am. Chem. Soc., **86**,
736 (1964)

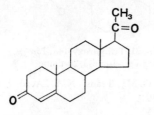

Progesterone

Calonectria decora

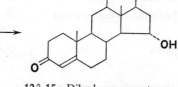

12β, 15α-Dihydroxyprogesterone

Ger. Pat. 1,067,020

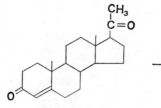

Progesterone

Calonectria decora (80%)

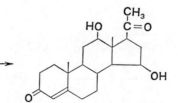

12β, 15β-Dihydroxyprogesterone

Schubert, A., G. Langbein and R. Siebert, Chem.
Ber., **90**, 2576 (1957)

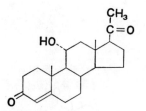

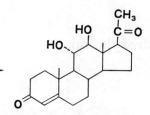

11α-Hydroxyprogesterone

Calonectria decora

11α, 12β-Dihydroxyprogesterone

Schubert, A., G. Langbein and R. Siebert, Chem. Ber., **90**, 2576 (1957)

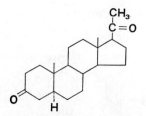

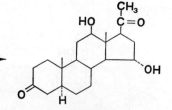

5α-Pregnane-3, 20-dione

Calonectria decora

12β, 15α-Dihydroxy-5α-pregnane-3, 20-dione

Ger. Pat. 1,067,020

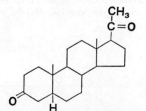

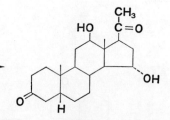

5β-Pregnane-3, 20-dione

Calonectria decora

12β, 15α-Dihydroxy-5β-pregnane-3, 20-dione

Ger. Pat. 1,067,020

(j) 14-Hydroxylation

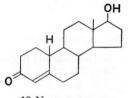

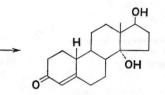

19-Nortestosterone

Mucor griseocyanus

14α-Hydroxy-19-nortestosterone

U. S. Pat. 2,662,089

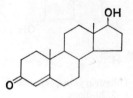

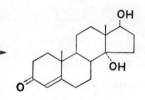

Testosterone

Mucor griseocyanus (ca. 100%)

Mucor sp. (35%)

14α-Hydroxytestosterone

Meister, P. D., S. H. Eppstein, D. H. Peterson, H. C. Murray, H. M. Leigh, A. Weintraub and L. M. Reineke, Abstr. paper 123rd Meeting Am. Chem. Soc., Los Angeles, March, 1953 p. 5c

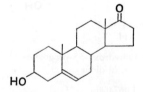

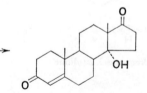

Dehydroepiandrosterone

Bacillus pulvifaciens IAM N-19-2

14α-Hydroxyandrost-4-ene-
3, 17-dione

Iizuka H., A. Naito and Y. Sato, J. Gen. Appl. Microbiol. (Japan), **7**, 118 (1961)

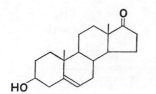
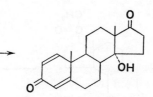

Dehydroepiandrosterone — 14α-Hydroxyandrosta-1, 4-
diene-3, 17-dione

Bacillus pulvifaciens IAM N-19-2   Iizuka, H., A. Naito and Y. Sato, J. Gen. Appl. Microbiol. (Japan), **7**, 118 (1961)

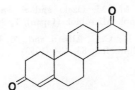
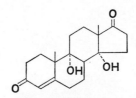

Androst-4-ene-3, 17-dione — 9α, 14α-Dihydroxyandrost-
4-ene-3, 17-dione

Cercospora melonis   Kondo, E. and K. Tori, J. Am. Chem. Soc., **86**, 736 (1964)

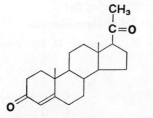
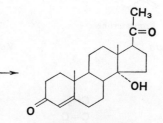

Progesterone — 14α-Hydroxyprogesterone

Absidia regnieri   Shirasaka, M., Chem. Pharm. Bull. (Japan), **9**, 59 (1961)

Bacillus cereus   Eppstein, S. H., P. D. Meister, H. C. Murray and D. H. Peterson, Vitamins and Hormones, **14**, 359 (1956)

Helicostylum piriforme   U. S. Pat. 2,670,358

Mucor griseocyanus (15%)

Mucor griseocyanus, sporangiospore   Vézina, C., S. N. Sehgal and K. Singh, Appl. Microbiol., **11**, 50 (1963)

Mucor parasiticus (15%)   U. S. Pat. 2,670,358

Stachylidium theobromae, conidia   Vézina, C., S. N. Sehgal and K. Singh, Appl. Microbiol., **11**, 50 (1963)

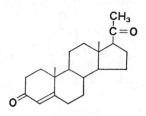 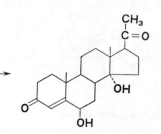

Progesterone   6β, 14α-Dihydroxyprogesterone

Achromobacter kashiwazakiensis IAM
K-40-5

Tsuda, K., H. Iizuka, E. Ohki, Y. Sato, A. Naito
and M. Hattori, J. Gen. Appl. Microbiol. (Japan), **5**, 7 (1959)

Bacillus cereus

Shirasaka, M., M. Ozaki and S. Sugawara, J.
Gem. Appl. Microbiol. (Japan), **7**, 341 (1961)

Mucor corymbifer

Camerino, B., C. G. Alberti and A. Vercellone,
Gazz. Chim. Ital, **83**, 684 (1953)

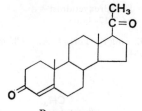 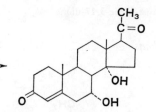

Progesterone   7α, 14α-Dihydroxyprogesterone

Curvularia lunata

Zetsche, K., Naturwiss., **47**, 232 (1960)

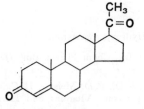 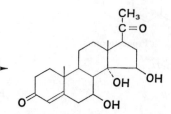

Progesterone   7β, 14α, 15β-Trihydroxyprogesterone

Syncephalastrum racemosum

Tsuda, K., T. Asai, Y. Sato, T. Tanaka, T.
Matsuhisa and H. Hasegawa, Chem. Pharm.
Bull. (Japan), **8**, 626 (1960)

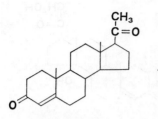
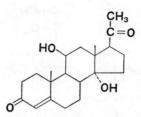

Progesterone

11β, 14α-Dihydroxyprogesterone

Curvularia lunata

Zetsche, K., Naturwiss., **47**, 232 (1960)

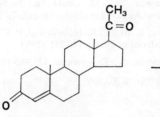
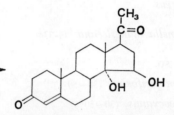

Progesterone

14α, 15β-Dihydroxyprogesterone

Helminthosporium sativum

Tsuda, K., T. Asai, Y. Sato, T. Tanaka and H. Hasegawa, Chem. Pharm. Bull. (Japan), **9**, 735 (1961)

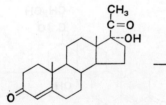
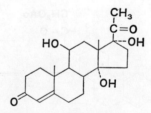

17α-Hydroxyprogesterone

11β, 14α, 17α-Trihydroxyprogesterone

Curvularia lunata NRRL 2380 (20%)

Shull, G. M. and D. A. Kita, J. Am. Chem. Soc., **77**, 763 (1955)

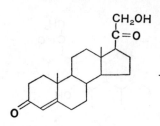 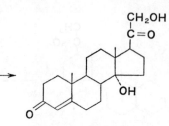

11-Deoxycorticosterone	14α, 21-Dihydroxypregn-4-ene-3, 20-dione

Absidia regnieri

Shirasaka, M., Chem. Pharm. Bull. (Japan), **9**, 59 (1961)

Bacillus cereus

Shirasaka, M., M. Ozaki and S. Sugawara, J. Gen. Appl. Microbiol. (Japan), **7**, 341 (1961)

Cunninghamella blakesleeana H-334

Mann, K. M., F. R. Hanson and P. W. O'Connell, Federation Proc., **14**, 251 (1955)

Curvularia sp.

Wettstein, A., Experientia, **11**, 465 (1955)

Helicostylum piriforme (30~50%)
Mucor griseocyanus (30~50%)

U. S. Pat. 2,703,806

Mucor parasiticus

Tamm, Ch., A. Gubler, G. Juhasz, E. Weiss-Berg and W. Zürcher, Helv. Chim. Acta., **46**, 889 (1963)

Stachylidium bicolor

Shirasaka, M., Chem. Pharm. Bull. (Japan), **9**, 203 (1961)

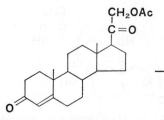 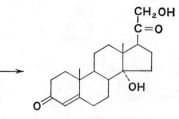

11-Deoxycorticosterone 21-acetate	14α, 21-Dihydroxypregn-4-ene-3, 20-dione

Mucor griseocyanus (25%)

Meister, P. D., S. H. Eppstein, D. H. Peterson, H. C. Murray, H. M. Leigh, A. Weintraub and L. M. Reineke, Abstr. 123rd Meeting Am. Chem. Soc., Los Angeles, March, 1953 p. 5c

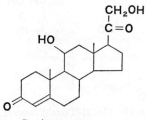

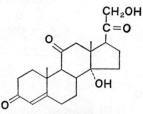

Corticosterone

14α, 21-Dihydroxypregn-4-
ene-3, 11, 20-trione

Absidia regnieri

Shirasaka, M., Chem. Pharm. Bull. (Japan), **9**, 59
(1961)

Bacillus cereus

Shirasaka, M., M. Ozaki and S. Sugawara, J. Gen.
Appl. Microbiol. (Japan), **7**, 341 (1961)

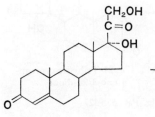

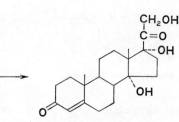

11-Deoxycortisol

14α, 17α, 21-Trihydroxypregn-
4-ene-3, 20-dione

Absidia regnieri

Shirasaka, M., Chem. Pharm. Bull. (Japan), **9**, 59
(1961)

Cunninghamella blakesleeana
Helicostylum piriforme

Eppstein, S. H., P. D. Meister, H. C. Murray and
D. H. Peterson, Vitamins and Hormones, **14**,
359 (1956)

Helminthosporium avenae H-9

Kondo, E., J. Agr. Chem. Soc. (Japan), **34**, 762
(1960)

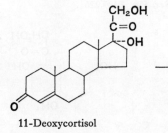

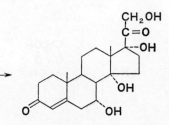

11-Deoxycortisol

7α, 14α, 17α, 21-Tetrahydroxypregn-
4-ene-3, 20-dione

Curvularia lunata

Shull, G. M., Trans. N. Y. Acad. Sci., **19**, 147
(1956)

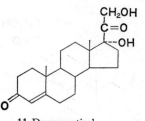

11-Deoxycortisol

Curvularia lunata NRRL 2380

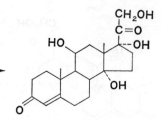

11β, 14α, 17α, 21-Tetrahydroxypregn-
4-ene-3, 20-dione

Agnello, E. J., B. L. Bloom and G. D. Laubach,
J. Am. Chem. Soc., **77**, 4684 (1955)

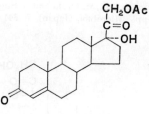

11-Deoxycortisol 21-acetate

Mycobacterium lacticola
Mycobacterium smegmatis

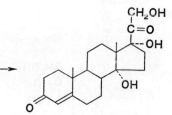

14α, 17α, 21-Trihydroxypregn-
4-ene-3, 20-dione

Belg. Pat. 538,327

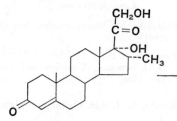

16α-Methyl-11-deoxycortisol

Curvularia lunata

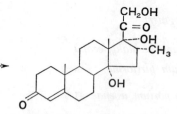

16α-Methyl-14α, 17α, 21-trihydroxy-
pregn-4-ene-3, 20-dione

Ger. Pat. 1,147,226

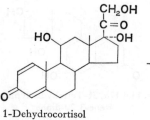

1-Dehydrocortisol
(Prednisolone)

Helicostylum piriforme

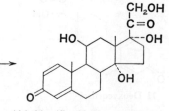

11β, 14α, 17α, 21-Tetrahydroxy-
pregna-1, 4-diene-3, 20-dione

Brit. Pat. 835,700

(k) 15-Hydroxylation

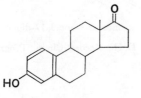

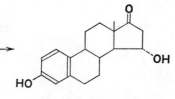

Estrone	15α-Hydroxyestrone

Fusarium moniliforme

Crabbé, P. and C. Casas-Campillo, J. Org. Chem. **29**, 2731 (1964)

Fusarium moniliforme ATCC 9851 (20%)
Fusarium moniliforme ATCC 11161 (10%)
Fusarium moniliforme IH 4 (53.5%)

Casas-Campillo, C. and M. Bautista, Appl. Microbiol., **13**, 977, (1965)

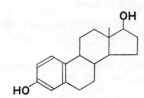

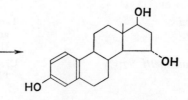

Estradiol	15α-Hydroxyestradiol

Fusarium moniliforme

Crabbé, P. and C. Casas-Campillo, J. Org. Chem., **29**, 2731 (1964)

Fusarium moniliforme ATCC 9851 (29%)
Fusarium moniliforme IH 4 (42.5%)

Casas-Campillo, C. and M. Bautista, Appl. Microbiol., **13**, 977 (1965)

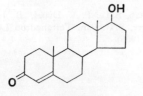

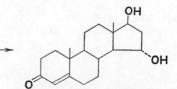

Testosterone	15α-Hydroxytestosterone

Fusarium lini

Tamm, Ch., A. Gubler, G. Juhasz, E. Weiss-Berg and W. Zürcher, Helv. Chim. Acta, **46**, 889, (1963)

Fusarium roseum

Rao, P. G., Indian J. Pharm., **25**, 131 (1963)

Gibberella saubinetti

Japan Pat. 313,770

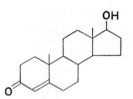

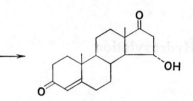

Testosterone

15α-Hydroxyandrost-4-ene-3, 17-dione

Fusarium sp.

Peterson, D. H., Record Chem. Progr., **17**, 211 (1956)

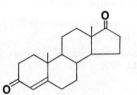

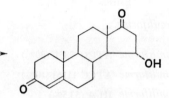

Androst-4-ene-3, 17-dione

15α-Hydroxyandrost-4-ene-3, 17-dione

Fusarium lini

Tamm, Ch., A. Gubler, G. Juhasz, E. Weiss-Berg and W. Zürcher, Helv. Chim. Acta., **46**, 889, (1963)

Gibberella saubinetti (21%)

Urech, J., E. Vischer and A. Wettstein, Helv. Chim. Acta, **43**, 1077 (1960)

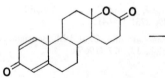

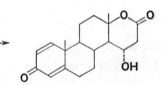

1-Dehydrotestololactone

15α-Hydroxy-1-dehydrotestololactone

Penicillium sp. ATCC 11598

Neidleman, S. L., P. A. Diassi, B. Junta. R. M. Palmere, S. C. Pan, Tetrahedron Letter No. 44 5337 (1966)

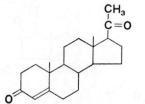 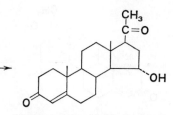

Progesterone   15α-Hydroxyprogesterone

Colletotrichum antirrhini	Eppstein, S. H., P. D. Meister, H. C. Murray and D. H. Peterson, Vitamins and Hormones, **14**, 359 (1956)
Fusarium culmorum	Klüger, B., R. Siebert and A. Schubert, Naturwiss, **44**, 40 (1957)
Fusarium lini	Tamm, Ch., A. Gubler, G. Juhasz, E. Weiss-Berg and W. Zürcher, Helv. Chim. Acta, **46**, 889 (1963)
Fusarium lycopersici	Shirasaka, M. and M. Tsuruta, Chem. Pharm. Bull. (Japan), **9**, 238 (1961)
Fusarium lycopersici	Klüger, B., R. Siebert and A. Schubert, Naturwiss, **44**, 40 (1957)
Fusarium roseum	Rao, P. G., Indian J. Pharm., **25**, 131 (1962)
Fusarium solani	Klüger, B., R. Siebert and A. Schubert, Naturwiss., **44**, 40 (1957)
Gibberella saubinetti	Shirasaka, M. and M. Tsuruta, Chem. Pharm. Bull. (Japan), **9**, 238 (1961)
Helminthosporium sativum	Tsuda, K., T. Asai, Y. Sato, T. Tanaka and H. Hasegawa, Chem. Pharm. Bull. (Japan), **9**, 735 (1961)
Nigrospora oryzae	U. S. Pat. 2,793,163
Penicillium notatum	Camerino, B., R. Modelli and C. Spalla, Gazz. Chim. Ital., **86**, 1226 (1956)
Penicillium urticae	Eppstein, S. H., P. D. Meiter, H. C. Murray and D. H. Peterson, Vitamins and Hormones, **14**, 359 (1956)
Streptomyces aureus	U. S. Pat. 2,753,290

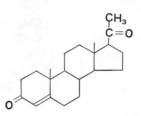

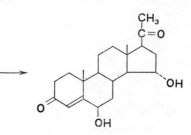

Progesterone   6β, 15α-Dihydroxyprogesterone

Fusarium lini   Tamm, Ch., A. Gubler, G. Juhasz, E. Weiss-Berg
                                and W. Zürcher, Helv. Chim. Acta, **46**, 889
                                (1963)

Fusarium roseum   Rao, P. G., Indian J. Pharm, **25**, 131 (1962)

Gibberella saubinetti   Japan Pat. 313,770

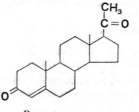

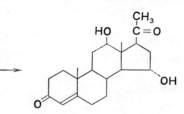

Progesterone   11α, 15α-Dihydroxyprogesterone

Nigrospora oryzae   U. S. Pat. 2,793,163

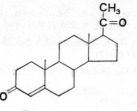

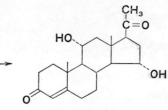

Progesterone   12β, 15α-Dihydroxyprogesterone

Calonectria decora   Ger. Pat. 1,067,020

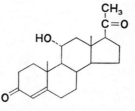

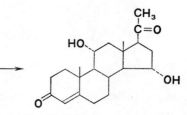

11α-Hydroxyprogesterone   11α, 15α-Dihydroxyprogesterone

Calonectria decora   Ger. Pat. 1,067,020

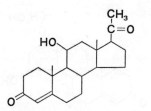

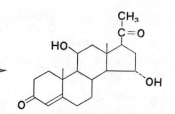

11β-Hydroxyprogesterone

Calonectria decora

11β, 15α-Dihydroxyprogesterone

Ger. Pat. 1,067,020

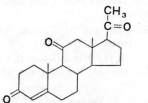

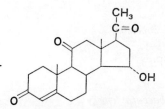

11-Oxoprogesterone

Calonectria decora

15α-Hydroxy-11-oxoprogesterone
(15α-Hydroxypregn-4-ene-3, 11, 20-trione)

Ger. Pat. 1,067,020

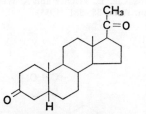

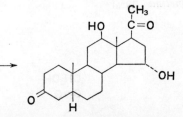

5α-Pregnane-3, 20-dione

Calonectria decora

12β, 15α-Dihydroxy-5α-pregnane-
3, 20-dione

Ger. Pat. 1,067,020

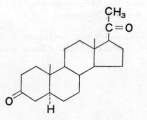

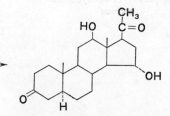

5β-Pregnane-3, 20-dione

Calonectria decora

12β, 15α-Dihydroxy-5β-pregnane-
3, 20-dione

Ger. Pat. 1,067,020

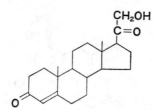

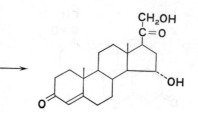

11-Deoxycorticosterone	15α, 21-Dihydroxypregn-4-ene-3, 20-dione

Fusarium lini

Tamm, Ch., A. Gubler, G. Juhasz, E. Weiss-Berg and W. Zürcher, Helv. Chim. Acta, **46**, 889, 1166 (1963)

Fusarium lycopersici

Shirasaka, M. and M. Tsuruta, Chem. Pharm. Bull. (Japan), **9**, 238 (1961)

Fusarium sp.

Eppstein, S. H., P. D. Meister, H. C. Murray and D. H. Peterson, Vitamins and Hormones, **14**, 359 (1956)

Gibberella baccata (70%)

Urech, J., E. Vischer and A. Wettstein, Helv. Chim. Acta, **43**, 1077 (1960)

Gibberella saubinetti

Shirasaka, M. and M. Tsuruta, Chem. Pharm. Bull. (Japan), **9**, 238 (1961)

Gibberella saubinetti (50%)

Urech, J., E. Vischer and A. Wettstein, Helv. Chim. Acta, **43**, 1077 (1960)

Lenzites abietina (19%)

Meystre, Ch., E. Vischer and A. Wettstein, Helv. Chim. Acta, **38**, 381 (1955)

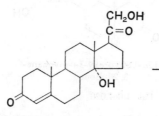

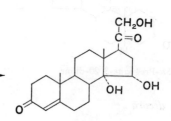

14α-Hydroxy-11-deoxycortico-sterone	14α, 15α, 21-Trihydroxypregn-4-ene-3, 20-dione

Fusarium lini

Tamm, Ch., A. Gubler, G. Juhasz, E. Weiss-Berg and W. Zürcher, Helv. Chim. Acta, **46**, 889 (1963)

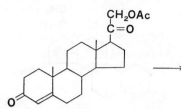 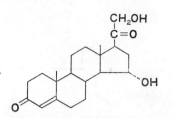

11-Deoxycorticosterone 21-acetate	15α, 21-Dihydroxypregn-4-ene-3, 20-dione
Calonectria decora	Ger. Pat. 1,067,020
Nigrospora oryzae	U. S. Pat. 2,793,163

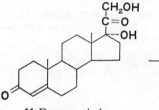 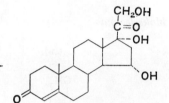

11-Deoxycortisol

15α, 17α, 21-Trihydroxypregn-
4-ene-3, 20-dione

Fusarium lini

Tamm, Ch., A. Gubler, G. Juhasz, E. Weiss-Berg and W. Zürcher, Helv. Chim. Acta, **46**, 889 (1963)

Fusarium roseum

Rao, P. G., Indian J. Pharm., **25**, 131 (1963)

Gibberella baccata (2~4%)

Gibberella saubinetti (5%)

Urech, J., E. Vischer and A. Wettstein, Helv. Chim. Acta, **43**, 1077 (1960)

Helminthosporium sativum (42%)

Tsuda, K., T. Asai, Y. Sato and T. Tanaka, Chem. Pharm. Bull. (Japan), **7**, 534 (1959)

Hormodendrum olivaceum

U. S. Pat. 3,010,877

Hormodendrum viride strain Lederle No. Z-10

Bernstein, S., L. I. Feldman, W. S. Allen, R. H. Blank and C. E. Linden, Chem. & Ind. 111 (1956)

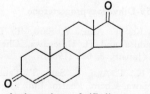 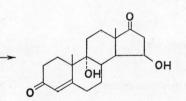

Androst-4-ene-3, 17-dione

9α, 15β-Dihydroxyandrost-
4-ene-3, 17-dione

Cercospora melonis

Kondo, E. and K. Tori, J. Am. Chem. Soc., **86**, 736 (1964)

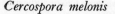

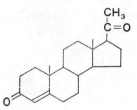

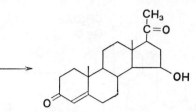

Progesterone   15β-Hydroxyprogesterone

Bacillus megaterium   NRRL B-938   McAleer, W. J., T. A. Jacob, L. B. Turnbull, E. F. Schoenewaldt and T. H. Stoudt, Arch. Biochem. Biophys., **73**, 127 (1958)

Helminthosporium sativum   Tsuda, K., T. Asai, Y. Sato, T. Tanaka and H. Hasegawa, Chem. Pharm, Bull. (Japan), **9**, 735 (1961)

Phycomyces blakesleeana   Eppstein, S. H., P. D., Meister, H. C. Murray and D. H. Peterson, Vitamins and Hormones, **14**, 359 (1956)

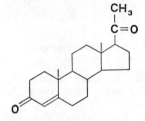

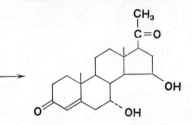

Progesterone   7α, 15β-Dihydroxyprogesterone

Helminthosporium sativum   Tsuda, K., T. Asai, Y. Sato, T. Tanaka and H. Hasegawa, Chem. Pharm. Bull (Japan), **9**, 735 (1961)

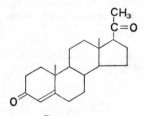

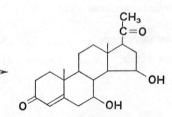

Progesterone   7β, 15β-Dihydroxyprogesterone

Diplodia tubericola   Tsuda, K., T. Asai, Y. Sato, T. Tanaka, T. Matsuhisa and H. Hasegawa, Chem. Pharm. Bull. (Japan), **8**, 626 (1960)

Helminthosporium sativum   Tsuda, K., T. Asai, Y. Sato, T. Tanaka and H. Hasegawa, Chem. Pharm. Bull. (Japan), **9**, 375 (1961)

Syncephalastrum racemosum   Tsuda, K., T. Asai, Y. Sato, T. Tanaka, and T. Matsuhisa and H. Hasegawa, Chem. Pharm. Bull. (Japan), **8**, 626 (1960)

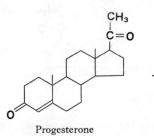

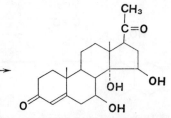

Progesterone

7β, 14α, 15β-Trihydroxyprogesterone

Syncephalastrum racemosum

Tsuda, K., T. Asai, Y. Sato, T. Tanaka, T. Matsuhisa and H. Hasegawa, Chem. Pharm. Bull. (Japan), **8**, 626 (1960)

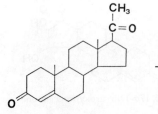

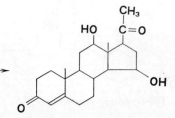

Progesterone

12β, 15β-Dihydroxyprogesterone

Calonectria decora (80%)

Schubert, A., G. Langbein and R. Siebert, Chem. Ber., **90**, 2576 (1957)

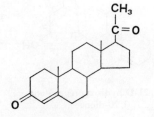

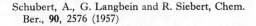

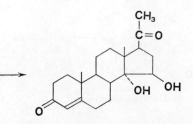

Progesterone

14α, 15β-Dihydroxyprogesterone

Helminthosporium sativum

Tsuda, K., T. Asai, Y. Sato, T. Tanaka and H. Hasegawa, Chem. Pharm. Bull. (Japan), **9**, 735 (1961)

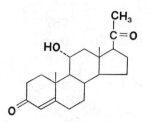
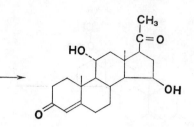

11α-Hydroxyprogesterone

Calonectria decora

11α, 15β-Dihydroxyprogesterone

Schubert, A., G. Langbein and R. Siebert, Chem. Ber., **90**, 2576 (1957)

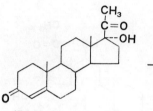
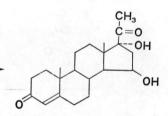

17α-Hydroxyprogesterone

Nigrospora oryzae

15β, 17α-Dihydroxyprogesterone

U. S. Pat. 2,793,163

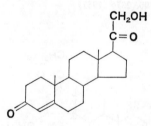
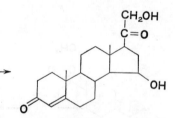

11-Deoxycorticosterone

15β, 21-Dihydroxypregn-4-ene-
3, 20-dione

Fusarium sp.

Eppstein, S. H., P. D. Meister, H. C. Murray and D. H. Peterson, Vitamins and Hormones, **14**, 359 (1956)

Gibberella baccata (20~60%)

Meystre, Ch., E. Vischer and A. Wettstein, Helv. Chim. Acta, **38**, 381 (1955)

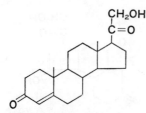

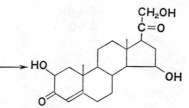

11-Deoxycorticosterone

2β, 15β, 21-Trihydroxypregn-
4-ene-3, 20-dione

Sclerotinia libertiana

Shirasaka, M., Chem. Pharm. Bull. (Japan), **9**, 54
(1961)

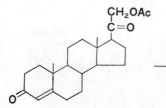

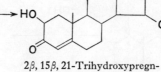

11-Deoxycorticosterone 21-acetate

2β, 15β, 21-Trihydroxypregn-
4-ene-3, 20-dione

Sclerotinia sclerotiorum

Japan Pat. 311,627

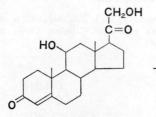

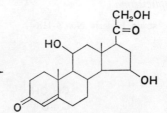

Corticosterone

11β, 15β, 21-Trihydroxypregn-
4-ene-3, 20-dione

Botrytis cinerea

Shirasaka, M., Chem. Pharm. Bull. (Japan), **9**, 152
(1961)

Sclerotinia libertiana

Shirasaka, M., Chem. Pharm. Bull. (Japan), **9**, 54
(1961)

Sclerotinia sclerotiorum

Japan Pat. 311,627

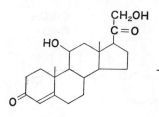

Corticosterone

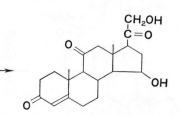

15β, 21-Dihydroxypregn-4-ene-
3, 11, 20-trione

Botrytis cinerea

Shirasaka, M., Chem. Pharm. Bull. (Japan), **9**, 152
(1961)

Sclerotium hydrophilum

Shirasaka, M. and M. Tsuruta, Chem. Pharm.
Bull. (Japan), **9**, 196 (1961)

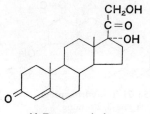

11-Deoxycortisol

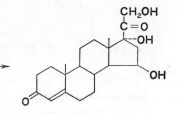

15β, 17α, 21-Trihydroxypregn-
4-ene-3, 20-dione

Bacillus megaterium

U. S. Pat. 2,958,631

Spicaria simplicissima

U. S. Pat. 3,010,877

Spicaria sp. strain Lederle No. Z-118

Bernstein, S., L. I. Feldman, W. S. Allen, R. H.
Blank and C. E. Linden, Chem. & Ind., 111
(1956)

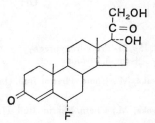

6α-Fluoro-11-deoxycortisol

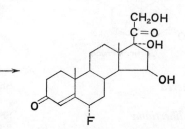

6α-Fluoro-15β, 17α, 21-trihydroxypregn-
4-ene-3, 20-dione

Aspergillus nidulans

U. S. Pat. 3,004,047

(1) 16-Hydroxylation

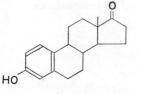 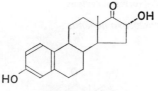

Estrone	16α-Hydroxyestrone

Streptomyces halstedii ATCC 13499, NRRL B-2138

Streptomyces mediocidicus ATCC 13278

Kita, D. A., J. L. Sardinas and G. M. Shull, Nature, **190**, 627 (1961)

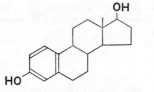 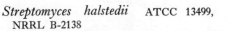 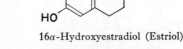

Estradiol	16α-Hydroxyestradiol (Estriol)

Streptomyces halstedii ATCC 13499, NRRL B-2138

Streptomyces mediocidicus ATCC 13278

Kita, D. A., J. L. Sardinas and G. M. Shull, Nature, **190**, 627 (1961)

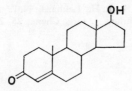 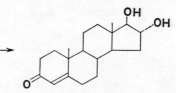

Testosterone	16α-Hydroxytestosterone

Pestalotia funera

Staurophoma sp.

Streptomyces roseochromogenus

Thoma, R. W. et al., 69th Meeting N. Y. C. Branch of Soc. Am. Bact. N. Y. (1955)

U. S. Pat. 3,071,516

Eppstein, S. H., P. D. Meister, H. C. Murray and D. H. Peterson, Vitamins and Hormones, **14**, 359 (1956)

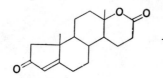

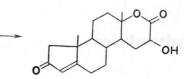

A-Nortestololactone

16α-Hydroxy-A-nortestololactone

Streptomyces roseochromogenus

U. S. Pat. 3,098,079

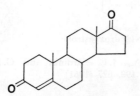

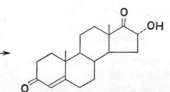

Androst-4-ene-3, 17-dione

16α-Hydroxyandrost-4-ene-3, 17-dione

Staurophoma sp.

U. S. Pat. 3,071,516

Streptomyces roseochromogenus

Eppstein, S. H., P. D. Meister, H. C. Murray and
D. H. Peterson, Vitamins and Hormones, 14, 359
(1956)

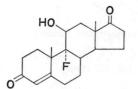

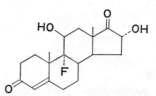

9α-Fluoro-11β-hydroxyandrost-
4-ene-3, 17-dione

9α-Fluoro-11β, 16α-dihydroxyandrost-
4-ene-3, 17-dione

Streptomyces roseochromogenus strain
Lederle AE-409

Bernstein, S., R. H. Lenhard, N. E. Rigler and
M. A. Darken, J. Org. Chem., 25, 297 (1960)

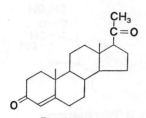
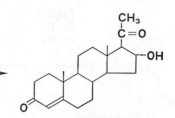

Progesterone

16α-Hydroxyprogesterone

Actinomyces sp. ATCC 11009

Perlman, D., E. Titus and J. Fried, J. Am. Chem. Soc., **74**, 2126 (1952)

Actinomycetes

Vondrová, O. and A. Čapek, Folia Microbiol, **8**, 117 (1963)

Staurophoma sp.

U. S. Pat. 3,071,516

Streptomyces sp. (30~40%)

Perlman, D., E. O'Brien, A. P. Bayan and R. B. Greenfield, J. Bacteriol, **69**, 347 (1955)

Streptomyces sp.

Shirasaka, M. and M. Ozaki, J. Ferm. Assoc. (Japan), **19**, 389 (1961)

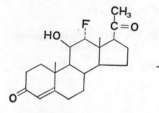
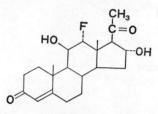

12α-Fluoro-11β-hydroxy-
progesterone

12α-Fluoro-11β, 16α-dihydroxy-
progesterone

Streptomyces roseochromogenus

Brit. Pat. 916,790

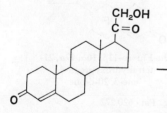
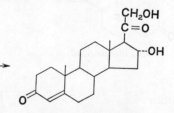

11-Deoxycorticosterone

16α, 21-Dihydroxypregn-
4-ene-3, 20-dione

Didymella vodakii

Wettstein, A., Experientia, **11**, 465 (1955)

Streptomyces roseochromogenus

Eppstein, S. H., P. D. Meister, H. C. Murray and D. H. Peterson, Vitamins and Hormones, **14**, 359 (1956)

Streptomyces sp. (15%)

Vischer, E., J. Schmidlin and A. Wettstein, Helv. Chim. Acta, **37**, 321 (1954)

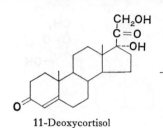

11-Deoxycortisol

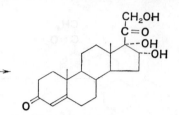

16α, 17α, 21-Trihydroxypregn-
4-ene-3, 20-dione

Streptomyces roseochromogenus, conidia

Streptomyces viridis, conidia

Vézina, C., S. N. Sehgal and K. Singh, Appl.
Microbiol., **11**, 50 (1963)

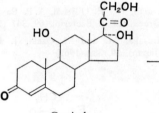

Cortisol

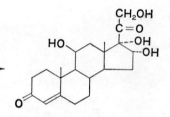

11β, 16α, 17α, 21-Tetrahydroxypregn-
4-ene-3, 20-dione

Nocardia italica

Belg. Pat. 620,272

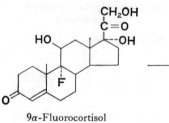

9α-Fluorocortisol

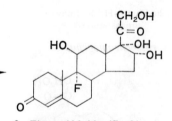

9α-Fluoro-11β, 16α, 17α, 21-
tetrahydroxypregn-4-
ene-3, 20-dione

Nocardia italica

Belg. Pat. 620,272

Streptomyces halstedii

U. S. Pat. 2,991,230

Streptomyces roseochromogenus strain
Waksman No. 3689 (50%)

Thoma, R. W., J. Fried, S. Bonanno and P.
Grabowich, J. Am. Chem. Soc., **79**, 4818 (1957)

Streptomyces roseochromogenus, spore

U. S. Pat. 2,982,693

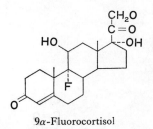

9α-Fluorocortisol

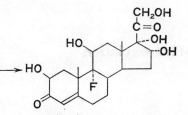

9α-Fluoro-2β, 11β, 16α, 17α, 21-
pentahydroxypregn-4-
ene-3, 20-dione

Streptomyces roseochromogenus   ATCC
3347 (71%)

Goodman, J. J. and L. L. Smith, Appl. Microbiol.,
9, 372 (1961)

Streptomyces roseochromogenus   strain
Waksman No. 3689 (75%)

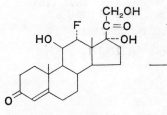

12α-Fluorocortisol

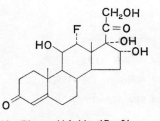

12α-Fluoro-11β, 16α, 17α, 21-
tetrahydroxypregn-4-
ene-3, 20-dione

Streptomyces roseochromogenus

Brit. Pat. 916,790

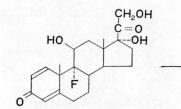

9α-Fluoroprednisolone

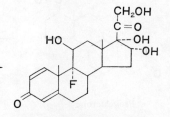

9α-Fluoro-11β, 16α, 17α, 21-
tetrahydroxypregna-1, 4-
diene-3, 20-dione

Nocardia italica

Belg. Pat. 620,272

Streptomyces roseochromogenus   strain
Waksman No. 3689 (20%)

Thoma, R. W., J. Fried, S. Bonanno and P.
Grabowich, J. Am. Chem. Soc., **79**, 4818 (1957)

(m) 17-Hydroxylation

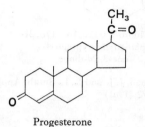 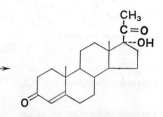

Progesterone   17α-Hydroxyprogesterone

Cephalothecium roseum

Meister, P. D., L. M. Reineke, R. C. Meeks, H. C. Murray, S. H. Eppstein, H. M. Leigh, A. Weintraub and D. H. Peterson, J. Am. Chem. Soc., **76**, 4050 (1954)

Sporormia minima

U. S. Pat. 2,813,060

Trichoderma album (15%)

Japan Pat. 228,171

Trichoderma koningi (10~15%)

Trichoderma sp. (10~15%)

Trichoderma sp. (*Trichoderma ligno-rum*)

Brit. Pat. 759,731

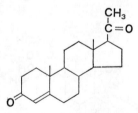 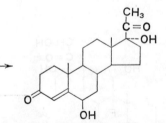

Progesterone   6β, 17α-Dihydroxyprogesterone

Naucoria confragosa C-172

Schuytema, E. C., M. P. Hargie, D. J. Siehr, I. Merits, J. R. Schenck, M. S. Smith and E. L. Varner, Appl. Microbiol., **11**, 256 (1963)

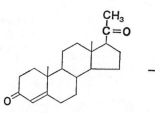

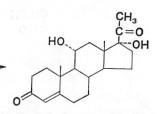

Progesterone

11α, 17α-Dihydroxyprogesterone

Cephalothecium roseum ATCC 8685

Meister, P. D., L. M. Reineke, R. C. Meeks, H. C. Murray, S. H. Eppstein, H. M. Leigh, A. Weintraub and D. H. Peterson, J. Am. Chem. Soc., **76**, 4050 (1954)

Dactylium dendroides

Dulaney, E. L., W. J. McAleer, H. R. Barkemeyer and C. Hlavac, Appl. Microbiol., **3**, 372 (1955)

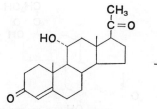

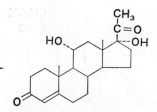

11α-Hydroxyprogesterone

11α, 17α-Dihydroxyprogesterone

Dactylium dendroides

Dulaney, E. L., W. J. McAleer, H. R. Barkemeyer and C. Hlavac, Appl. Microbiol., **3**, 372 (1955)

Sepedonium ampullosporum

U. S. Pat. 3,011,951

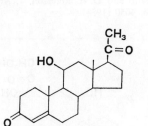

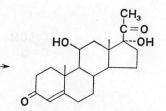

11β-Hydroxyprogesterone

11β, 17α-Dihydroxyprogesterone

Sporormia minima

U. S. Pat. 2,813,060

Trichoderma viride (8~10%)

Japan Pat. 228,171

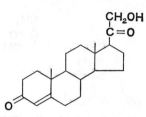

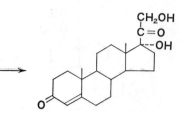

11-Deoxycorticosterone

17α, 21-Dihydroxypregn-4-ene-
3, 20-dione (11-Deoxycortisol)

Sporormia minima

U. S. Pat. 2,813,060

Trichothecium roseum

Meystre Ch., E. Vischer and A. Wettstein, Helv.
Chim. Acta, **37**, 1548 (1954)

Trichoderma viride (8~10%)

Japan Pat. 228,171

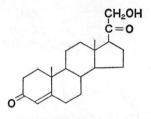

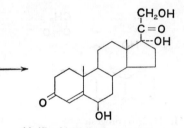

11-Deoxycorticosterone

6β, 17α, 21-Trihydroxypregn-
4-ene-3, 20-dione

Cephalothecium roseum ATCC 8685

Meister, P.D., L. M. Reineke, R. C. Meeks, H. C.
Murray, S. H., Eppstein, H. M. Leigh, A. Weintraub and D. H. Peterson, J. Am. Chem. Soc.,
76, 4050 (1954)

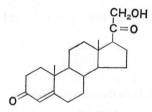

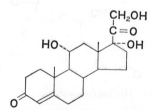

11-Deoxycorticosterone

11α, 17α, 21-Trihydroxypregn-
4-ene-3, 20-dione

Cephalothecium roseum ATCC 8685

Meister, P. D., L. M. Reineke, R. C. Meeks, H. C.
Murray, S. H. Eppstein, H. M. Leigh, A. Weintraub and D. H. Peterson, J. Am. Chem. Soc.,
76, 4050 (1954)

Scopulariopsis brevicaulis

U. S. Pat. 2,970,085

Trichothecium roseum

— 90 —

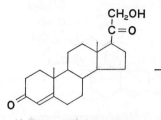

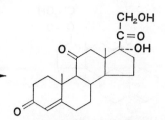

11-Deoxycorticosterone

17α, 21-Dihydroxypregn-4-ene-
3, 11, 20-trione (Cortisone)

Cephalothecium roseum ATCC 8685

Meister, P. D., L. M. Reineke, R. C. Meeks, H. C.
Murray, S. H. Eppstein, H. M. Leigh, A. Weintraub and D. H. Peterson, J. Am. Chem. Soc.,
76, 4050 (1954)

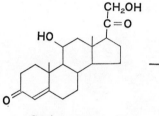

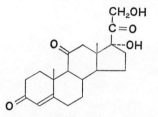

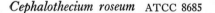

Corticosterone

17α, 21-Dihydroxypregn-4-ene-
3, 11, 20-trione (Cortisone)

Cephalothecium roseum ATCC 8685

Meister, P. D., L. M. Reineke, R. C. Meeks, H. C.
Murray, S. H. Eppstein, H. M. Leigh, A. Weintraub and D. H. Peterson, J. Am. Chem. Soc.,
76, 4050 (1954)

Trichothecium roseum

Meystre Ch., E. Vischer and A. Wettstein, Helv.
Chim. Acta, **37**, 1548 (1954)

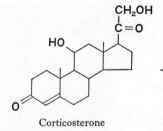

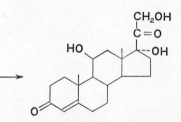

Corticosterone

11β, 17α, 21-Trihydroxypregn-4-ene-
3, 20-dione (Cortisol)

Cephalothecium roseum ATCC 8685

Meister, P. D., L. M. Reineke, R. C. Meeks, H. C.
Murray, S. H. Eppstein, H. M. Leigh, A. Weintraub and D. H. Peterson, J. Am. Chem. Soc.,
76, 4050 (1954)

Sporormia minima

U. S. Pat. 2,813,060

Trichothecium roseum

Meystre, Ch., E. Vischer and A. Wettstein, Helv.
Chim. Acta, **37**, 1548 (1954)

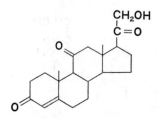

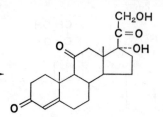

11-Dehydrocorticosterone

Trichothecium roseum

17α, 21-Dihydroxypregn-4-ene-
3, 11, 20-trione (Cortisone)

Meystre, Ch., E. Vischer and A. Wettstein, Helv..
Chim. Acta, **37**, 1548 (1954)

(n) 18-Hydroxylation

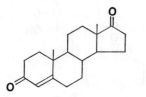 →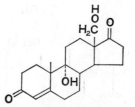

Androst-4-ene-3, 17-dione

9α, 18-Dihydroxyandrost-
4-ene-3, 17-dione

Cercospora melonis

Kondo, E. and K. Tori, J. Am. Chem. Soc., **86**,
736 (1964)

(o) 19-Hydroxylation

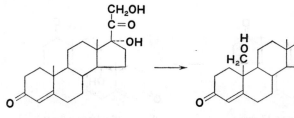

11-Deoxycortisol	17α, 19, 21-Trihydroxypregn- 4-ene-3, 20-dione
Corticium sasakii	Hasegawa, T. and T. Takahashi, Bull. Agr. Chem. Soc. (Japan), **22**, 212 (1958)
Corticium vagum	Japan Pat. 276,077
Pestalotia sp.	Japan Pat. 310,308

(p) 21-Hydroxylation

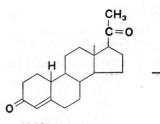

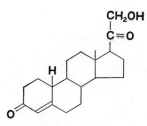

19-Norprogesterone

21-Hydroxy-19-norpregn-
4-ene-3, 20-dione

Aspergillus niger ATCC 9142

Zaffaroni, A., C. Casas-Campillo, F. Cordoba and
G. Rosenkranz, Experientia, **11**, 219 (1955)

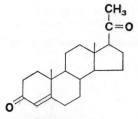

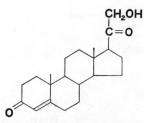

Progesterone

21-Hydroxypregn-4-ene-3, 20-dione
(11-Deoxycorticosterone)

Aspergillus niger ATCC 9142

Zaffaroni, A., C. Casas-Campillo, F. Cordoba and
G. Rosenkranz, Experientia, **11**, 219 (1955)

Cercosporella herpotrichoides

U. S. Pat. 3,056,730

Hendersonia sp. (25~30%)

Japan Pat. 228,172

Kabatiella phoradendri

U. S. Pat. 2,977,286

Ophiobolus herbotrichus (60%)

Meystre, Ch., E. Vischer and A. Wettstein, Helv.
Chim. Acta, **37**, 1548 (1954)

Sclerotinia fructicola

U. S. Pat. 2,778,776

Wojnowicia graminis (35~45%)

McAleer, W. J. and E. L. Dulaney, Arch. Biochem.
Biophys., **62**, 109 (1956)

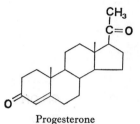

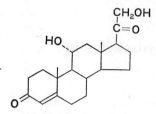

Progesterone

11α, 21-Dihydroxypregn-
4-ene-3, 20-dione

Aspergillus sp.

Weisz, E., G. Wix and M. Bodánszky, Naturwiss.,
43, 39 (1956)

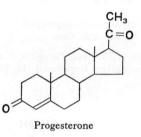

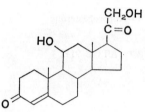

Progesterone

11β, 21-Dihydroxypregn-4-ene-
3, 20-dione (Corticosterone)

Curvularia lunata

Rubin, B. A. et al., Bact. Proc., **56**, 33 (1956)

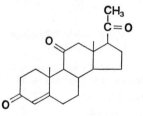

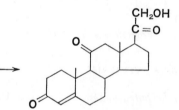

11-Oxoprogesterone

21-Hydroxypregn-4-ene-3, 11, 20-trione

Aspergillus niger ATCC 9142

Zaffaroni, A., C. Casas-Campillo, F. Cordoba and
G. Rosenkranz, Experientia, **11**, 219 (1955)

Ophiobolus herbotrichus

Meystre, Ch., E. Vischer and A. Wettstein, Helv.
Chim. Acta, **37**, 1548 (1954)

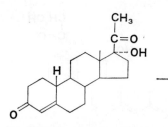
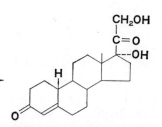

17α-Hydroxy-19-norprogesterone

17α, 21-Dihydroxy-19-norpregn-
4-ene-3, 20-dione

Ophiobolus herbotrichus

Zaffaroni, A., H. J. Ringold, G. Rosenkranz, F. Sondheimer, G. H. Thomas and C. Djerassi, J. Am. Chem. Soc., **76**, 6210 (1954)

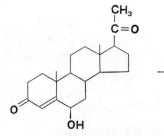
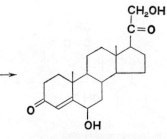

6β-Hydroxyprogesterone

6β, 21-Dihydroxypregn-
4-ene-3, 20-dione

Aspergillus niger ATCC 9142

Zaffaroni, A., C. Casas-Campillo, F. Cordoba and G. Rosenkranz, Experientia, **11**, 219 (1955)

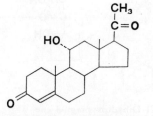
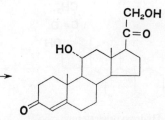

11α-Hydroxyprogesterone

11α, 21-Dihydroxypregn-4-ene-3, 20-
dione (11-Epicorticosterone)

Aspergillus niger ATCC 9142

Hendersonia sp. (10%)

Zaffaroni, A., C. Casas-Campillo, F. Cordoba and G. Rosenkranz, Experientia, **11**, 219 (1955)

Japan Pat. 228,172

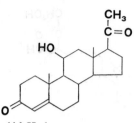

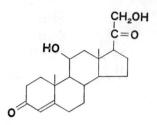

11β-Hydroxyprogesterone

11β, 21-Dihydroxypregn-4-ene-
3, 20-dione (Corticosterone)

Aspergillus niger ATCC 9142

Hendersonia acicola (10~15%)

Zaffaroni, A., C. Casas-Campillo, F. Cordoba and
G. Rosenkranz, Experientia, **11**, 219 (1955)

Brit. Pat. 767,360

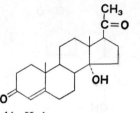

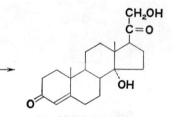

14α-Hydroxyprogesterone

14α, 21-Dihydroxypregn-
4-ene-3, 20-dione

Aspergillus niger ATCC 9142

Zaffaroni, A., C. Casas-Campillo, F. Cordoba and
G. Rosenkranz, Experientia, **11**, 219 (1955)

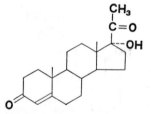

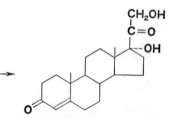

17α-Hydroxyprogesterone

17α, 21-Dihydroxypregn-4-ene-
3, 20-dione (11-Deoxycortisol)

Aspergillus niger (18%)

Ophiobolus herbotrichus

Brit. Pat. 749,943

Meystre, Ch., E. Vischer and A. Wettstein, Helv.
Chim. Acta, **37**, 1548 (1954)

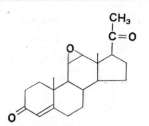
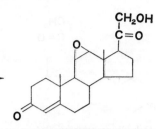

11β, 12β-Oxidoprogesterone

Cercosporella herpotrichoides

11β, 12β-Oxido-21-hydroxypregn-
4-ene-3, 20-dione

U. S. Pat. 3,056,730

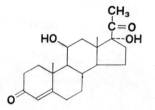
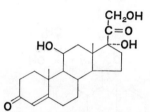

11β, 17α-Dihydroxyprogesterone

Hendersonia herpotricia (5%)

11β, 17α, 21-Trihydroxypregn-4-ene-
3, 20-dione (Cortisol)

Japan Pat. 228,172

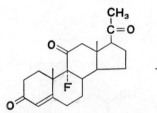
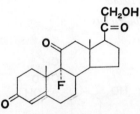

9α-Fluoro-11-oxoprogesterone

Cercosporella herpotrichoides

9α-Fluoro-21-hydroxypregn-
4-ene-3, 11, 20-trione

U. S. Pat. 3,056,730

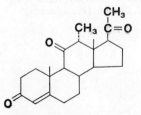
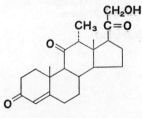

12α-Methyl-11-oxoprogesterone

Cercosporella herpotrichoides
Kabatiella phoradendri

12α-Methyl-21-hydroxypregn-4-
ene-3, 11, 20-trione

U. S. Pat. 3,056,730

U. S. Pat. 2,977,286

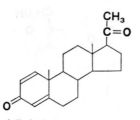
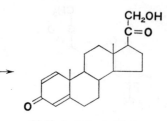

1-Dehydroprogesterone

21-Hydroxypregna-1, 4-
diene-3, 20-dione

Ophiobolus herbotrichus

U. S. Pat. 2,778,776

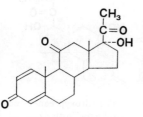
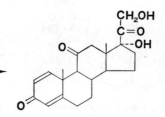

17α-Hydroxypregna-1, 4-
diene-3, 11, 20-trione

17α, 21-Dihydroxypregna-1, 4-diene-
3, 11, 20-trione (Prednisone)

Sclerotinia fructicola
Ophiobolus herbotrichus

U. S. Pat. 2,778,776

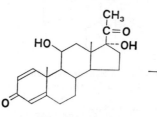
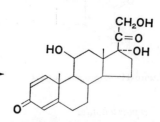

11β, 17α-Dihydroxypregna-1, 4-
diene-3, 20-dione

11β, 17α, 21-Trihydroxypregna-1, 4-
diene-3, 20-dione (Prednisolone)

Ophiobolus herbotrichus

U. S. Pat. 2,778,776

B. Dehydrogenation

(a) $\rangle$CH–OH $\longrightarrow$ $\rangle$C=O

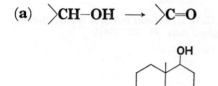

Estradiol

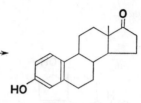

Estrone

Actinomycetes of the Albus group (100%)   Welsch, M. and C. Heusghem, Compt. Rend. Soc. Biol., **142**, 1074 (1948)

Flavobacterium dehydrogenans   Ercoli, A., Biochim. Terap. sper., **28**, 125 (1941)

Proactinomyces sp.   Turfitt, G. E., Biochem. J., **42**, 376 (1948)

Pseudodiphtheria bacilli   Zimmermann, W. and G. May, Zntr. Bakt. Parasitenk. I Abt, **151**, 462 (1944)

Pseudomonas testosteroni ATCC 11996   Hurlock, B. and P. Talalay, J. Biol. Chem., **233**, 886 (1958)

Streptomyces diastaticus, conidia   Vézina, C., S. N. Sehgal and K. Singh, Appl. Microbiol, **11**, 50 (1963)

Streptomyces rimosus, conidia

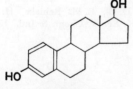

Estradiol

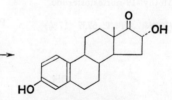

16α-Hydroxyestrone

Streptomyces halstedii ATCC 13499   Kita, D. A., J. L. Sardinas and G. M. Shull, Nature, **190**, 627 (1961)
Streptomyces mediocidicus ATCC 13278

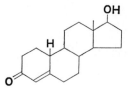

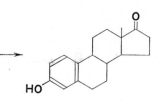

19-Nortestosterone

Estrone

Bacillus sphaericus ATCC 7055

Gaul, C., R. I. Dorfman and S. R. Stitch, Biochem.
Biophys. Acta, **49**, 387 (1961)

Nocardia corallina

U. S. Pat. 3,087,864

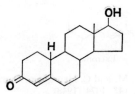

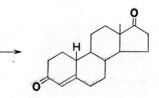

19-Nortestosterone

19-Norandrost-4-ene-3, 17-dione

Bacillus sphaericus ATCC 7055

Gaul, C., R. I. Dorfman and S. R. Stitch, Biochem.
Biophys. Acta, **49**, 387 (1961)

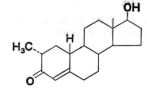

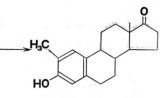

2α-Methyl-19-nortestosterone

2-Methylestrone

Septomyxa affinis ATCC 6737 (17%)

Peterson, D. H., L. M. Reineke, H. C. Murray
and O. K. Sebek, Chem. & Ind., 1301 (1960)

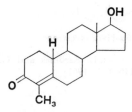

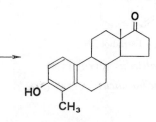

4-Methyl-19-nortestosterone

4-Methylestrone

Septomyxa affinis ATCC 6737 (17%)

Peterson, D. H., L. M. Reineke, H. C. Murray
and O. K. Sebek, Chem. & Ind., 1301 (1960)

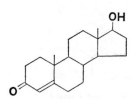

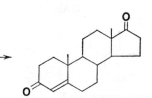

Testosterone

Androst-4-ene-3, 17-dione

Bacillus pulvifaciens IAM N-19-2

Iizuka, H., A. Naito and Y. Sato, J. Gen. Appl. Microbiol. (Japan), **7**, 118 (1961)

Pseudomonas chlororaphis IAM 1511

Naito, A., Y. Sato, H. Iizuka and K. Tsuda, Steroids, **3**, 327 (1964)

Pseudomonas testosteroni ATCC 11996

Marcus, P. I. and P. Talalay, J. Biol. Chem., **218**, 661 (1956)

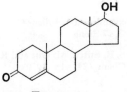

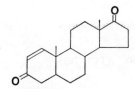

Testosterone

Androst-1-ene-3, 17-dione

Nocardia corallina

U. S. Pat. 3,087,864

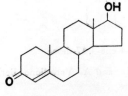

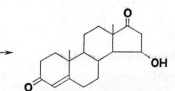

Testosterone

15α-Hydroxyandrost-4-ene-3, 17-dione

Fusarium sp.

Peterson, D. H., Record Chem. Progr., **17**, 211 (1956)

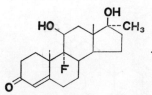

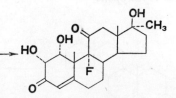

9α-Fluoro-17α-methyl-11β, 17β-dihydroxyandrost-4-en-3-one

9α-Fluoro-17α-methyl-1α, 2α, 17β-tri-hydroxyandrost-4-ene-3, 11-dione

Nocardia corallina ATCC 999 (18%)

Sax, K. J., C. E. Holmlund, L. I. Feldman, R. H. Evans, Jr., R. H. Blank, A. J. Shay, J. S. Schultz and M. Dann, Steroids, **5**, 345 (1965)

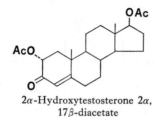

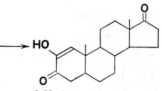

2α-Hydroxytestosterone 2α,
17β-diacetate

2-Hydroxyandrost-1-ene-3,
17-dione

Nocardia corallina

U. S. Pat. 3,087,864

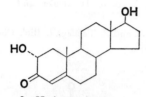

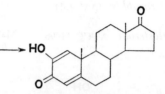

2α-Hydroxytestosterone

2-Hydroxyandrosta-1, 4-diene-
3, 17-dione

Bacillus sphaericus   ATCC 7055

Gaul, C., S. R. Stitch, M. Gut and R. I. Dorfman,
J. Org. Chem., **24**, 418 (1959)

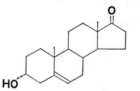

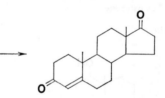

3α-Hydroxyandrost-5-en-17-one

Androst-4-ene-3, 17-dione

Aerobic bacteria (87%)

Mamoli, L. and A. Vercellone Ber., **71B**, 1686
(1938)

Flavobacterium carbonilicum

Molina, L., A. Ercoli, Boll. ist sieroterap milanese,
23, 164 (1944)

Flavobacterium dehydrogenans

Arnaudi, C., Boll. sez. ital., Soc. intern. Microbiol.,
11, 208 (1939)

Proactinomyces erythropolis

Turfitt, G. E., Biochem. J, **40**, 79 (1946)

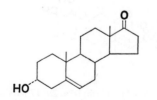

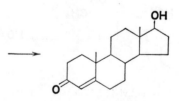

3α-Hydroxyandrost-5-en-17-one

Testosterone

Impoverished yeast

U. S. Pat. 2,236,574

Oxidizing bacteria (81%)

Mamoli, L., Ber., **71B**, 2278 (1938)

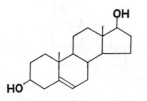

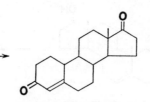

Androst-5-ene-3β, 17β-diol

Androst-4-ene-3, 17-dione

Flavobacterium androstenedionicum
Flavobacterium carbonilicum

Molina, L. and A. Ercoli, Boll. ist sieroterap milanese, **23**, 164 (1944)

Flavobacterium dehydrogenans (4%)
Proactinomyces erythropolis

Ercoli, A., Z. physiol. Chem., **270**, 266 (1941)

Turfitt, G. E., Biochem. J., **40**, 79 (1946)

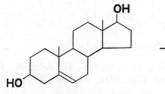

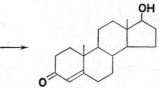

Androst-5-ene-3β, 17β-diol

Testosterone

Flavobacterium carbonilicum

Molina, L. and A. Ercoli, Boll. ist sieroterap. milanese, **23**, 164 (1944)

Flavobacterium dehydrogenans (64%)

Ercoli, A., Z. Physiol. Chem., **270**, 266 (1941)

Pseudodiphtheria bacilli

Zimmermann, W. and G. May, Zentr. Bakt. Parasitenk. IAbt., **151**, 462 (1944)

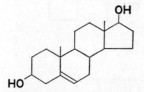

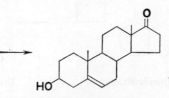

Androst-5-ene-3β, 17β-diol

3β-Hydroxyandrost-5-en-17-one

Pseudomonas sp.

Talalay, P. and M. M. Dobson, J. Biol. Chem, **205**, 823 (1953)

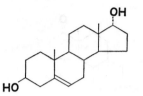

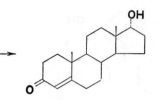

Androst-5-ene-3β, 17α-diol

17α-Hydroxyandrost-4-
en-3-one
(Epitestosterone)

Yeast

U. S. Pat. 2,186,906

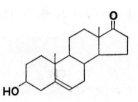

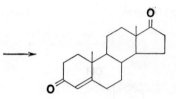

Dehydroepiandrosterone

Androst-4-ene-3, 17-dione

Alcaligenes faecalis (100%)

Hughes, H. B. and L. H. Schmidt, Proc. Soc.
Expt. Biol. Med., **51**, 162 (1942)

Bacillus pulvifaciens IAM N-19-2

Iizuka, H., A. Naito and Y. Sato, J. Gen. Appl.
Microbiol (Japan), **7**, 118 (1961)

Fusarium caucasicum

Vischer, E. and A. Wettstein, Experientia, **9**, 371
(1953)

Fusarium solani

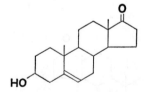

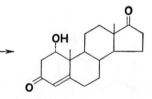

Dehydroepiandrosterone

1α-Hydroxyandrost-4-ene-
3, 17-dione

Penicillium sp.

Dodson, R. M., A. H. Goldkamp and R. D. Muir,
J. Am. Chem. Soc., **79**, 3921 (1957)

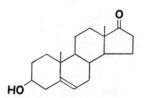

Dehydroepiandrosterone

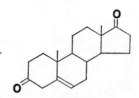

Androst-5-ene-3, 17-dione

Pseudomonas testosteroni ATCC 11996

Talalay, P. and P. I. Marcus, J. Biol. Chem., **218**, 675 (1956)

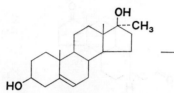

17α-Methylandrost-5-ene-
3β, 17β-diol

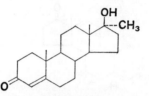

17α-Methyltestosterone

Dehydrogenating bacteria (75%)

Mamoli, L., Gazz. Chim. Ital., **69**, 237 (1939)

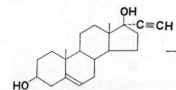

17α-Ethinylandrost-5-ene-
3β, 17β-diol

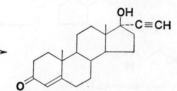

17α-Ethinyltestosterone

Flavobacterium dehydrogenans

Ercoli, A., Biochim. Terap. sper., **28**, 125 (1941)

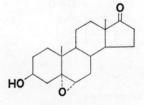

3β-Hydroxy-5α, 6α-oxido-
androstan-17-one

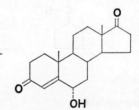

6α-Hydroxyandrost-4-ene-
3, 17-dione

Nocardia restrictus No. 545

Lee, S. S. and C. J. Sih, Biochemistry, **3**, 1267 (1964)

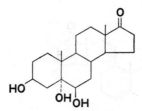

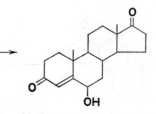

3β, 5α, 6β-Trihydroxy
androstan-17-one

6β-Hydroxyandrost-4-ene-
3, 17-dione

Nocardia restrictus No. 545

Lee, S.S. and C.J. Sih, Biochemistry, **3**, 1267
(1964)

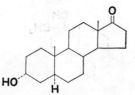

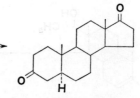

3α-Hydroxy-5α-androstan-17-one

5α-Androstane-3, 17-dione

Pseudomonas sp.

Talalay, P. and P.I. Marcus, Nature, **173**, 1189
(1954)

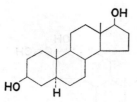

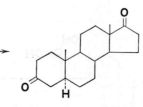

5α-Androstane-3β, 17β-diol

5α-Androstane-3, 17-dione

Flavobacterium dehydrogenans

Arnaudi, C., Experientia, **7**, 81 (1951)

Pseudomonas testosteroni ATCC 11996

Talalay, P. and P.I. Marcus, J. Biol. Chem, **218**,
675 (1956)

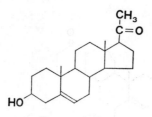

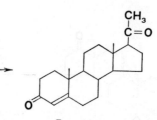

Pregnenolone

Progesterone

Aspergillus niger

Perlman, D., Science, **115**, 529 (1952)

Bacillus pulvifaciens IAM N-19-2

Iizuka, H., A. Naito and Y. Sato, J. Gen. Appl.
Microbiol (Japan), **7**, 118 (1961)

Bacterial mixture

Mamoli, L., Ber., **71 B**, 2701 (1938)

Flavobacterium dehydrogenans (82%)

Phycomyces blakesleeanus

Streptomyces sp.

Ercoli, A., Boll. Sci. Fac. Chim. Ind. Bologna, **279**
(1940)

Perlman, D., Science, **115**, 529 (1952)

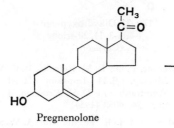

Pregnenolone

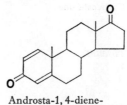

Androsta-1, 4-diene-
3, 17-dione

Fusarium caucasicum

Fusarium solani

Pycnodothis sp.

Vischer, E. and A. Wettstein, Experientia, **9**, 371
(1953)

Shull, G. M., Trans. N.Y. Acad. Sci., **19**, 147 (1956)

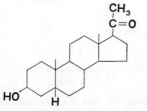

3β-Hydroxy-5β-pregnan-20-one

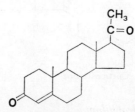

Progesterone

Streptomyces sp.

Perlman, D., E. O'Brien, A. P. Bayan and R. B.
Greenfield, Jr., J. Bacteriol, **69**, 347 (1955)

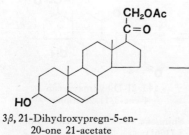

3β, 21-Dihydroxypregn-5-en-
20-one 21-acetate

21-Hydroxypregn-4-ene-
3, 20-dione
(11-Deoxycorticosterone)

Corynebacterium mediolanum (34%)

Mamoli, L., Ber., **72 B**, 1863 (1939)

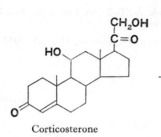

Corticosterone

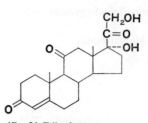

17α, 21-Dihydroxypregn-
4-ene-3, 11, 20-trione
(Cortisone)

Cephalothecium roseum

Meister, P. D., L. M. Reineke, R. C. Meeks, H. C. Murray, S. H. Eppstein, H. M. Leigh, A. Weintraub and D. H. Peterson, J. Am. Chem. Soc., **76**, 4050 (1954)

Trichothecium roseum

Meystre, Ch. E. Vischer and A. Wettstein, Helv. Chim. Acta, **37**, 1548 (1954)

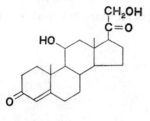

Corticosterone

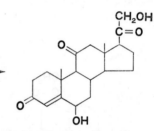

6β, 21-Dihydroxypregn-4-ene-
3, 11, 20-trione

Sclerotium hydrophilum

Shirasaka, M. and M. Tsuruta, Chem. Pharm. Bull (Japan), **9**, 196 (1961)

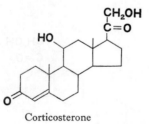

Corticosterone

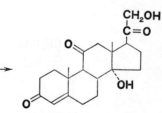

14α, 21-Dihydroxypregn-
4-ene-3, 11, 20-trione

Absidia regnieri

Shirasaka, M., Chem. Pharm. Bull. (Japan), **9**, 59 (1961)

Bacillus cereus

Shirasaka, M., M. Ozaki and S. Sugawara, J. Gen. Appl. Microbiol. (Japan), **7**, 341 (1961)

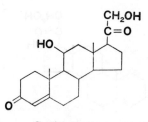

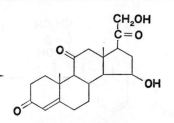

Corticosterone

15β, 21-Dihydroxypregn-
4-ene-3, 11, 20-trione

Botrytis cinerea

Shirasaka, M., Chem. Pharm. Bull. (Japan), **9**, 152
(1961)

Sclerotium hydrophilum

Shirasaka, M. and M. Tsuruta, Chem. Pharm.
Bull. (Japan), **9**, 196 (1961)

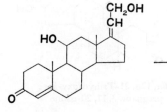

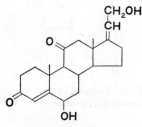

11β, 21-Dihydroxy-cis-pregna-
4, 17(20)-dien-3-one

6β, 21-Dihydroxy-cis-pregna-
4, 17(20)-diene-3, 11-dione

Rhizopus arrhizus

Hanze, A. R., O. K. Sebek and H. C. Murray, J.
Org. Chem., **25**, 1968 (1960)

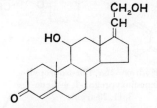

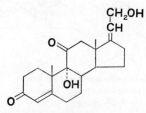

11β, 21-Dihydroxy-cis-pregna-
4, 17(20)-dien-3-one

9α, 21-Dihydroxy-cis-pregna-
4, 17(20)-diene-3, 11-dione

Helicostylum piriforme
Cunninghamella blakesleeana

Hanze, A. R., O. K. Sebek and H. C. Murray, J.
Org. Chem., **25**, 1968 (1960)

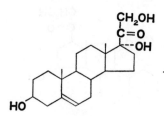 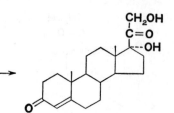

3β, 17α, 21-Trihydroxy-
pregn-5-en-20-one

17α, 21-Dihydroxypregn-4-ene-
3, 20-dione
(11-Deoxycortisol)

Corynebacterium mediolanum

U. S. Pat. 3,030,278

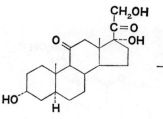 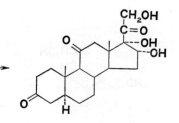

3α, 17α, 21-Trihydroxy-5α-
pregnane-11, 20-dione

16α, 17α, 21-Trihydroxy-5α-
pregnane-3, 11, 20-trione

Pseudomonas sp.

Talalay, P. and P. I. Marcus, Nature, **173**, 1189
(1954)

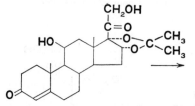 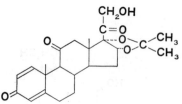

11β, 21-Dihydroxy-16α, 17α-iso-
propylidenedioxypregn-4-ene-
3, 20-dione

21-Hydroxy-16α, 17α-isopropyli-
denedioxypregna-1, 4-diene-
3, 11, 20-trione

Nocardia corallina ATCC 999

Sax, K. J., C. E. Holmlund, L. I. Feldman, R. H.
Evans, Jr., R. H. Blank, A. J. Shay, J. S. Schultz
and M. Dann, Steroids, 5, 345 (1965)

(b) $\diagup$CH$_2$ $\longrightarrow$ $\diagup$C=O

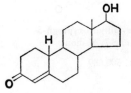

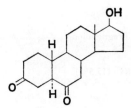

19-Nortestosterone

17β-Hydroxy-19-nor-5α-
androstane-3, 6-dione

Rhizopus reflexus

U. S. Pat. 2,692,273

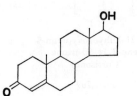

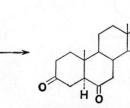

Testosterone

17β-Hydroxy-5α-andro-
stane-3, 6-dione

Rhizopus reflexus

U. S. Pat. 2,692,273

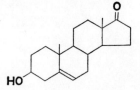

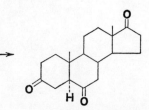

Dehydroepiandrosterone

5α-Androstane-3, 6, 17-trione

Bacillus pulvifaciens IAM N-19-2

Iizuka, H., A. Naito and Y. Sato, J. Gen. Appl.
Microbiol. (Japan), **7**, 118 (1961)

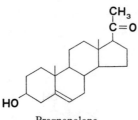

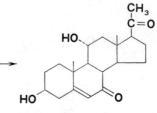

Pregnenolone

3β, 11α-Dihydroxypregn-5-
ene-7, 20-dione

Rhizopus arrhizus (17.5%)

Can. Pat. 506,689

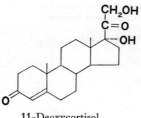

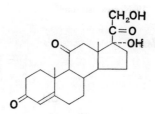

11-Deoxycortisol

17α, 21-Dihydroxypregn-4-
ene-3, 11, 20-trione
(Cortisone)

Cunninghamella   blakesleeana (9.5∼
13.0%)

U. S. Pat. 2,602,769

(c) $-CH_2-CH_2- \longrightarrow -CH=CH-$

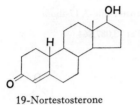

19-Nortestosterone

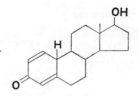

17β-Hydroxy-19-norandrosta-
1, 4-dien-3-one

Protaminobacter alboflavum

Protaminobacter rubrum

U. S. Pat. 2,776,927

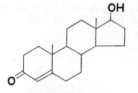

Testosterone

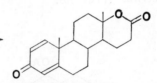

1-Dehydrotestololactone

Cylindrocarpon radicicola (50%)

Fried, J., R.W. Thoma and A. Klingsberg, J. Am. Chem. Soc., **75**, 5764 (1953)

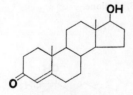

Testosterone

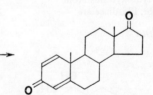

Androsta-1, 4-diene-3, 17-dione

Bacillus pulvifaciens IAM N-19-2

Pseudomonas chlororaphis IAM 1511

Pseudomonas testosteroni ATCC 11996

Iizuka, H., A. Naito and Y. Sato, J. Gen. Appl. Microbiol. (Japan), **7**, 118 (1961)

Naito, A., Y. Sato, H. Iizuka and K. Tsuda, Steroids, **3**, 327 (1964)

Levy, H. R. and P. Talalay, J. Am. Chem. Soc., **79**, 2658 (1957)

— 115 —

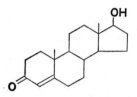 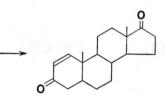

Testosterone

Androst-1-ene-3, 17-dione

Nocardia corallina

U. S. Pat. 3,087,864

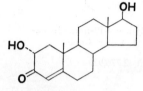 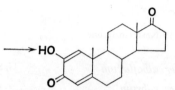

2α-Hydroxytestosterone

2-Hydroxyandrosta-1, 4-diene-
3, 17-dione

Bacillus sphaericus ATCC 7055

Gaul, C., S. R. Stitch, M. Gut and R. I. Dorfman,
J. Org. Chem., **24**, 418 (1959)

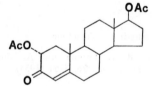 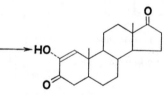

2α-Hydroxytestosterone
2α, 17β-diacetate

2-Hydroxyandrost-1-ene-
3, 17-dione

Nocardia corallina

U. S. Pat. 3,087,864

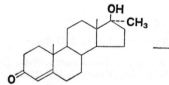 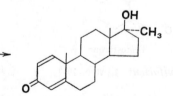

17α-Methyltestosterone

17α-Methyl-17β-hydroxyandrosta-
1, 4-dien-3-one

Bacterium aromaticus

Japan Pat. 289,328

Didymella lycopersici

Vischer, E., Ch. Meystre and A. Wettstein, Helv.
Chim. Acta., **38**, 1502 (1955)

Flavobacterium aquatile

Japan Pat. 305,166

Fusarium solani

Pseudomonas graveolens

Stereum fasciatum

Brit. Pat. 917,081

Japan Pat. 413,161

Japan Pat. 307,724

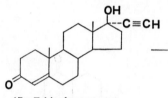 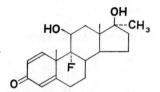

9α-Fluoro-17α-methyl-11β-
hydroxytestosterone

9α-Fluoro-17α-methyl-11β, 17β-
dihydroxyandrosta-1, 4-
dien-3-one

Nocardia corallina ATCC 999 (12%)

Sax, K. J., C. E. Holmlund, L. I. Feldman, R. H.
Evans, Jr., R. H. Blank, A. J. Shay, J. S. Schultz
and M. Dann, Steroids, **5**, 345 (1965)

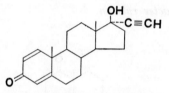

17α-Ethinyltestosterone

17α-Ethinyl-17β-hydroxyandrosta-
1, 4-dien-3-one

Didymella lycopersici

Vischer, E., Ch. Meystre and A. Wettstein, Helv.
Chim. Acta., **38**, 1502 (1955)

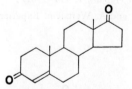 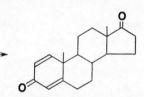

Androst-4-ene-3, 17-dione

Androsta-1, 4-diene-3, 17-dione

Bacillus pulvifaciens IAM N-19-2

Iizuka, H., A. Naito and Y. Sato, J. Gen. Appl.
Microbiol. (Japan), **7**, 118 (1961)

Bacillus sphaericus

Stoudt, T. H., W. J. McAleer, J. M. Chemerda,
M. A. Kozlowski, R. F. Hirschmann, V. Marlatt
and R. Miller, Arch. Biochem. Biophys, **59**, 304
(1955)

Fusarium caucasicum

Ger. Pat. 1,135,455

Fusarium lateritium (80%)

Čapek, A., O. Hanč and M. Tadra, Folia Microbiol,
8, 120 (1963)

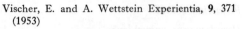

Fusarium solani (100%)   Vischer, E. and A. Wettstein Experientia, **9**, 371 (1953)

Nocardia restrictus   Sih, C. J. Biochem. Biophys. Res. Comm., **7**, 87 (1962)

Protaminobacter alboflavum   U.S. Pat. 2,776,928

Protaminobacter rubrum

Pseudomonas testosteroni ATCC 11996   Levy, H. R. and P. Talalay, J. Am. Chem. Soc., **79**, 2658 (1957)

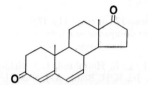   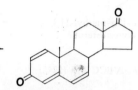

Androsta-4, 6-diene-3, 17-dione   Androsta-1, 4, 6-triene-3, 17-dione

Protaminobacter alboflavum   U.S. Pat. 2,776,927

Protaminobacter rubrum

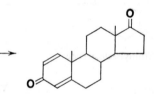

Dehydroepiandrosterone   Androsta-1, 4-diene-3, 17-dione

Bacillus pulvifaciens IAM N-19-2   Iizuka, H., A. Naito and Y. Sato, J. Gen. Appl. Microbiol (Japan), **7**, 118 (1961)

Fusarium caucasicum   Vischer, E. and A. Wettstein, Experientia, **9**, 371 (1953)

Fusarium solani

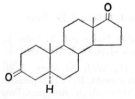   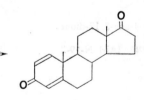

5α-Androstane-3, 17-dione   Androsta-1, 4-diene-3, 17-dione

Mycobacterium smegmatis   Brit. Pat. 850,951

Pseudomonas testosteroni ATCC 11996   Levy, H. R. and P. Talalay, J. Am. Chem. Soc., **79**, 2658 (1957)

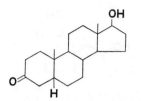
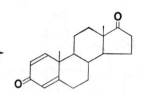

17β-Hydroxy-5β-androstan-3-one

Androsta-1, 4-diene-3, 17-dione

Pseudomonas testosteroni ATCC 11996

Levy, H. R. and P. Talalay, J. Am. Chem. Soc.,
79, 2658 (1957)

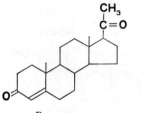
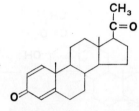

Progesterone

Pregna-1, 4-diene-3, 20-dione
(1-Dehydroprogesterone)

A Gram positive bacteria (90%)

Catroux, G. and H. Blachére, Ann. Inst. Pasteur,
105, 162 (1963)

Bacillus sphaericus

Stoudt, T. H., W. J. McAller, J. M. Chemerda,
M. A. Kozlowski, R. F. Hirschmann, V. Marlatt
and R. Miller, Arch. Biochem. Biophys, 59, 304
(1955)

Calonectria decora

Vischer, E., Ch. Meystre and A. Wettstein, Helv.
Chim. Acta, 38, 835 (1955)

Cylindrocarpon radicicola ATCC 11011

Peterson, G. E., R. W. Thoma, D. Perlman and
J. Fried, J. Bacteriol, 74, 684 (1957)

Didymella lycopersici

Vischer, E., Ch. Meystre and A. Wettstein, Helv.
Chim. Acta, 38, 1502 (1955)

Fusarium solani

Nishikawa, M., S. Noguchi and T. Hasegawa,
Pharm. Bull. (Japan), 3, 322 (1955)

Gloeosporium olivarum

Kondo, E. and E. Masuo, J. Agr. Chem. Soc.
(Japan), 34, 847 (1960)

Nocardia sp.

Sih, C. J. and R. E. Bennett, Biochem. Biophys.
Acta, 38, 378 (1960)

Septomyxa affinis ATCC 6737

Spero, G. B., J. L. Thompson, B. J. Magerlein,
A. R. Hanze, H. C. Murray, O. K. Sebek and
J. A. Hogg, J. Am. Chem. Soc., 78, 6213 (1956)

Septomyxa affinis, conidia

Vézina, C., S. N. Sehgal and K. Singh, Appl.
Microbiol., 11, 50 (1963)

Streptomyces lavendulae

Peterson, G. E., R. W. Thoma, D. Perlman and
J. Fried, J. Bacteriol., 74, 684 (1957)

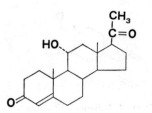 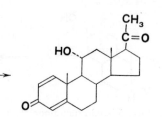

11α-Hydroxyprogesterone

11α-Hydroxypregna-1, 4-
diene-3, 20-dione

Bacillus cyclooxydans

U. S. Pat. 2,822,318

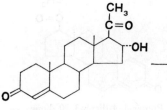 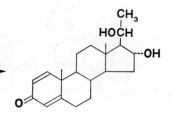

16α-Hydroxyprogesterone

16α, 20β-Dihydroxypregna-
1, 4-dien-3-one

Streptomyces lavendulae

Fried, J., R. W. Thoma, D. Perlman, J. E. Herz
and A. Borman, Recent Progr. Horm. Res., **11**,
149 (1955)

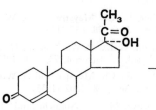 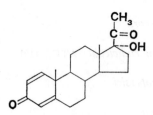

17α-Hydroxyprogesterone

17α-Hydroxypregna-1, 4-diene-
3, 20-dione

Protaminobacter alboflavum
Protaminobacter rubrum

U. S. Pat. 2,776,927

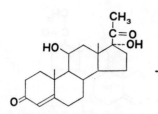

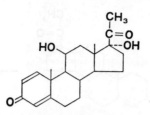

11β, 17α-Dihydroxyprogesterone

11β, 17α-Dihydroxypregna-1, 4-diene-3, 20-dione

Protaminobacter alboflavum

Protaminobacter rubrum

U.S. Pat. 2,776,927

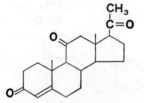

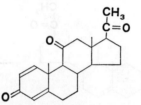

11-Oxoprogesterone

Pregna-1, 4-diene-3, 11, 20-trione

Protaminobacter alboflavum

Protaminobacter rubrum

U. S. Pat. 2,776,927

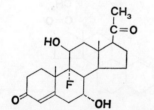

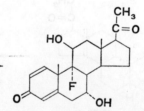

9α-Fluoro-7α, 11β-dihydroxy-progesterone

9α-Fluoro-7α, 11β-dihydroxy-pregna-1, 4-diene-3, 20-dione

Nocardia corallina

U. S. Pat. 2,962,512

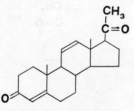

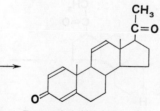

11-Dehydroprogesterone

Pregna-1, 4, 11-triene-3, 20-dione

Didymella lycopersici

Vischer, E., Ch. Meystre and A. Wettstein, Helv. Chim. Acta, **38**, 1502 (1955)

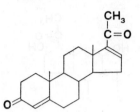

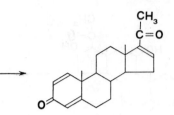

16-Dehydroprogesterone

Protaminobacter alboflavum

Protaminobacter rubrum

Pregna-1, 4, 16-triene-3, 20-dione

U. S. 2,776,927

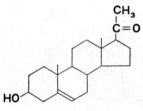

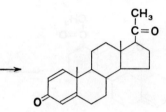

Pregnenolone

Bacillus pulvifaciens IAM N-19-2

Pregna-1, 4-diene-3, 20-dione

Iizuka, H., A. Naito and Y. Sato, J. Gen. Appl. Microbiol (Japan), **7**, 118 (1961)

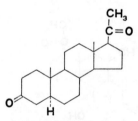

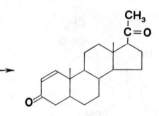

5α-Pregnane-3, 20-dione

Septomyxa affinis ATCC 6763

Pregn-1-ene-3, 20-dione

Fonken, G. S. and H. C. Murray, J. Org. Chem., **27**, 1102 (1962)

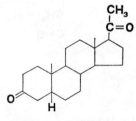

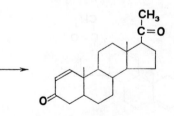

5β-Pregnane-3, 20-dione

Septomyxa affinis ATCC 6737

Pregn-1-ene-3, 20-dione

Fonken, G. S. and H. C. Murray, J. Org. Chem., **27**, 1102 (1962)

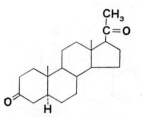

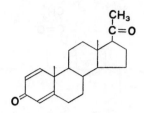

5α-Pregnane-3, 20-dione

Pregna-1, 4-diene-3, 20-dione

Nocardia sp.

Sih, C. J. and R. E. Bennett, Biochem. Biophys.
Acta, **38**, 378 (1960)

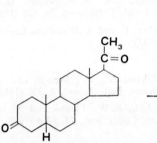

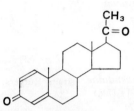

5β-Pregnane-3, 20-dione

Pregna-1, 4-diene-3, 20-dione

Nocardia sp.

Sih, C. J. and R. E. Bennett, Biochem. Biophys.
Acta, **38**, 378 (1960)

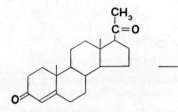

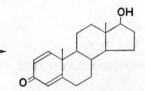

Progesterone

17β-Hydroxyandrosta-1, 4-dien-
3-one (1-Dehydrotestosterone)

Cylindrocarpon radicicola ATCC 11011

Fried, J., R. W. Thoma and A. Klingsberg, J.
Am. Chem. Soc., **75**, 5764 (1953)

Fusarium solani

Nishikawa, M., S. Noguchi and T. Hasegawa,
Pharm. Bull. (Japan), **3**, 322 (1955)

Streptomyces lavendulae strain Rutgers
Univ. No. 3440-14 (12%)

Fried, J., R. W. Thoma and A. Klingsberg, J.
Am. Chem. Soc., **75**, 5764 (1953)

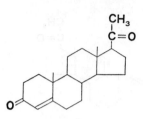
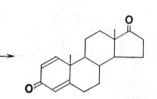

| Progesterone | Androsta-1, 4-diene-3, 17-dione |

Calonectria decora — Vischer, E., Ch. Meystre and A. Wettstein, Helv. Chim. Acta, **38**, 835 (1955)

Cylindrocarpon radicicola ATCC 11011 — Peterson, G. E., R. W. Thoma, D. Perlman and J. Fried, J. Bacteriol, **74**, 684 (1957)

Fusarium caucasicum
Fusarium solani — Vischer, E. and A. Wettstein, Experientia, **9**, 371 (1953)

Fusarium solani — Nishikawa, M., S. Noguchi and T. Hasegawa, Pharm. Bull. (Japan), **3**, 322 (1955)

Streptomyces lavendulae strain Rutgers Univ. No. 3440-14 (7%) — Fried, J., R. W. Thoma and A. Klingsberg, J. Am. Chem. Soc., **75**, 5764 (1953)

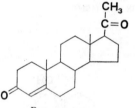
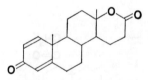

| Progesterone | 1-Dehydrotestololactone |

Cylindrocarpon radicicola ATCC 11011 (50%) — Fried, J., R. W. Thoma and A. Klingsberg, J. Am. Chem. Soc., **75**, 5764 (1953)

Fusarium lateritium (40%) — Čapek, A., O. Hanč and M. Tadra, Folia Microbiol., **8**, 120 (1963)

Fusarium solani — Nishikawa, M., S. Noguchi and T. Hasegawa, Pharm. Bull. (Japan), **3**, 322 (1955)

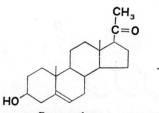

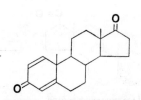

Pregnenolone | Androsta-1, 4-diene-3, 17-dione

Fusarium caucasicum
Fusarium solani

Vischer, E. and A. Wettstein, Experientia, **9**, 371 (1953)

Pycnodothis sp.

Shull, G. M., Trans. N. Y. Acad. Sci., **19**, 147 (1956)

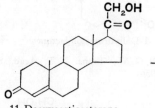

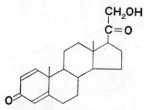

11-Deoxycorticosterone | 21-Hydroxypregna-1, 4-diene-3, 20-dione

Bacillus sphaericus

Stoudt, T. H., W. J. McAleer, J. M. Chemerda, M. A. Kozlowski, R. F. Hirschmann, V. Marlatt and R. Miller, Arch. Biochen. Biophys., **59**, 304 (1955)

Calonectria decora

Vischer, E., Ch. Meystre and A. Wettstein, Helv. Chim. Acta, **38**, 835 (1955)

Didymella lycopersici

Vischer, E., Ch. Meystre and A. Wettstein, Helv. Chim. Acta, **38**, 1502 (1955)

Gliocladium roseum
Helminthosporium turcicum
Ophiobolus heterostropus

Shirasaka, M. and M. Tsuruta, Chem. Pharm. Bull. (Japan), **9**, 207 (1961)

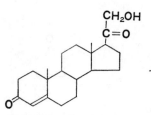

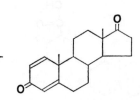

11-Deoxycorticosterone

Androsta-1, 4-diene-3, 17-dione

Fusarium caucasicum
Fusarium solani

Vischer, E. and A. Wettstein, Experientia, **9**, 371 (1953)

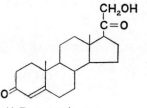

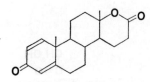

11-Deoxycorticosterone

1-Dehydrotestololactone

Fusarium sp.

Shull, G. M., Trans. N. Y. Acad. Sci., **19**, 147 (1956)

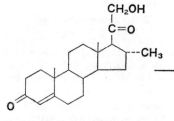

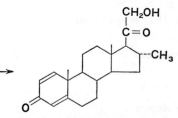

16α-Methyl-11-deoxycorticosterone

16α-Methyl-21-hydroxypregna-
1, 4-diene-3, 20-dione

Arthrobacter simplex
Bacillus lentus

Belg. Pat. 614,196

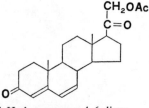

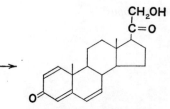

21-Hydroxypregna-4, 6-diene-
3, 20-dione 21-acetate

21-Hydroxypregna-1, 4, 6-
triene-3, 20-dione

Didymella lycopersici

Vischer, E., Ch. Meystre and A. Wettstein, Helv. Chim. Acta, **38**, 1502 (1955)

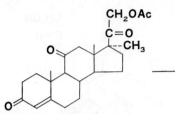

17α-Methyl-21-hydroxypregn-
4-ene-3, 11, 20-trione 21-acetate

Didymella lycopersici

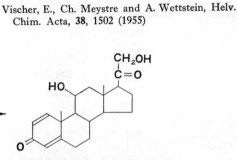

17α-Methyl-21-hydroxypregna-
1, 4-diene-3, 11, 20-trione

Vischer, E., Ch. Meystre and A. Wettstein, Helv.
Chim. Acta, **38**, 1502 (1955)

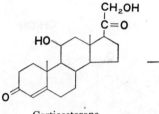

Corticosterone

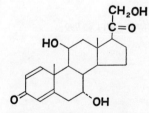

11β, 21-Dihydroxypregna-
1, 4-diene-3, 20-dione

Bacillus pulvifaciens IAM N-19-2

Iizuka, H., A. Naito and Y. Sato, J. Gen. Appl.
Microbiol. (Japan), **7**, 118 (1961)

Bacillus sphaericus

Stoudt, T. H., W. J. McAleer, J. M. Chemerda,
M. A. Kozlowski, R. F. Hirschmann, V. Marlatt
and R. Miller, Arch. Biochem. Biophys, **59**, 304
(1955)

Calonectria decora

Vischer, E., Ch. Meystre and A. Wettstein, Helv.
Chim. Acta, **38**, 835 (1955)

Didymella lycopersici

Vischer, E., Ch. Meystre and A. Wettstein, Helv.
Chim. Acta, **38**, 1502 (1955)

Gliocladium roseum

Shirasaka, M. and M. Tsuruta, Chem. Pharm.
Bull. (Japan), **9**, 207 (1961)

Helminthosporium turcicum

Ophiobolus heterostropus

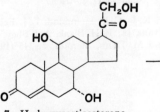

7α-Hydroxycorticosterone

7α, 11β, 21-Trihydroxypregna-
1, 4-diene-3, 20-dione

Nocardia corallina

U. S. Pat. 2,962,512

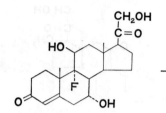

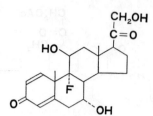

9α-Fluoro-7α-hydroxycorticosterone

9α-Fluoro-7α, 11β, 21-trihydroxy-
pregna-1, 4-diene-3, 20-dione

Nocardia corallina

U. S. Pat. 2,962,512

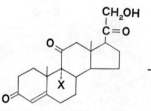

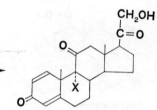

9α-Halo-11-dehydrocorticosterone

9α-Halo-21-hydroxypregna-1, 4-
diene-3, 11, 20-trione

Arthrobacter simplex
Corynebacterium hoagii

U. S. Pat. 3,084,103

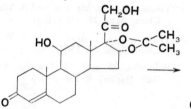

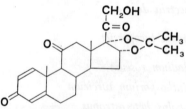

11β, 21-Dihydroxy-16α, 17α-isopro-
pylidenedioxypregn-4-ene-3, 20-dione

21-Hydroxy-16α, 17α-isopro-
pylidenedioxypregna-1, 4-
diene-3, 11, 20-trione

Nocardia corallina ATCC 999

Sax, K. J., C. E. Holmlund, L. I. Feldman, R. H.
Evans, Jr., R. H. Blank, A. J. Shay, J. S. Schultz
and M. Dann, Steroids, 5, 345 (1965)

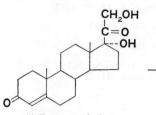 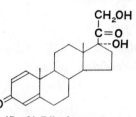

11-Deoxycortisol   17α, 21-Dihydroxypregna-
                                                 1, 4-diene-3, 20-dione

Bacillus cyclooxydans	U. S. Pat. 2,822,318
Bacillus pulvifaciens IAM N-19-2	Iizuka, H., A. Naito and Y. Sato, J. Gen. Appl. Microbiol. (Japan), **7**, 118 (1961)
Bacillus sphaericus	Stoudt, T. H., W. J. McAleer, J. M. Chemerda, M. A. Kozlowski, R. F. Hirschmann, V. Marlatt and R. Miller, Arch. Biochem. Biophys., **59**, 304 (1955)
Didymella lycopersici	Vischer, E., Ch. Meystre and A. Wettstein, Helv. Chim. Acta, **38**, 1502 (1955)
Flavobacterium aquatile	Japan Pat. 305,166
Fusarium solani	Vischer, E., Ch. Meystre and A. Wettstein, Helv. Chim. Acta, **38**, 835 (1955)
Fusarium solani, conidia	Vézina, C., S. N. Sehgal and K. Singh, Appl. Microbiol., **11**, 50 (1963)
Gliocladium roseum	Shirasaka, M. and M. Tsuruta, Chem. Pharm. Bull. (Japan), **9**, 207 (1961)
Gloeosporium olivarum	Kondo, E. and E. Masuo, J. Agr. Chem. Soc. (Japan), **34**, 847 (1960)
Helminthosporium gramineum	
Helminthosporium turcicum	Shirasaka, M. and M. Tsuruta, Chem. Pharm. Bull. (Japan), **9**, 207 (1961)
Helminthosporium zizaniae	Kondo, E. and E. Masuo, J. Agr. Chem. Soc. (Japan), **34**, 847 (1960)
Mixed culture of *Rhizopus nigricans* and *Bacillus subtilis* (80%)	Weisz, E., G. Wix and M. Bodánszky, Naturwiss., **43**, 39 (1956)
Mycobacterium lacticola ATCC 9626	Sutter, D., W. Charney, P. L. O'Neill, F. Carvajal, H. L. Herzog and E. B. Hershberg, J. Org. Chem. **22**, 578 (1957)
Ophiobolus heterostropus	Shirasaka, M. and M. Tsuruta, Chem. Pharm. Bull. (Japan), **9**, 207 (1961)
Protaminobacter alboflavum	U. S. Pat. 2,776,927
Protaminobacter rubrum	
Pseudomonas chlororaphis IAM 1511	Naito, A., Y. Sato, H. Iizuka and K. Tsuda, Steroids, **3**, 327 (1964)
Pseudomonas dacunhae	Shirasaka, M., M. Ozaki and S. Sugawara, J. Ferm. Assoc. (Japan), **19**, 335 (1961)

Septomyxa affinis, conidia

Serratia marcescens

Stereum fasciatum

Streptomyces lavendulae, conidia

Vézina, C., S. N. Sehgal and K. Singh, Appl. Microbiol., **11**, 50 (1963)

Japan Pat. 289,343

Japan Pat. 307,724

Vézina, C., S. N. Sehgal and K. Singh, Appl. Microbiol., **11**, 50 (1963)

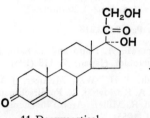

11-Deoxycortisol

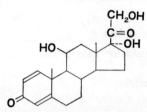

11β, 17α, 21-Trihydroxypregna-1, 4-diene-3, 20-dione (Prednisolone)

Absidia orchidis

Mixed culture of *Corticium sasakii* and *Pseudomonas boreopolis*

The actions of *Helminthosporium sativum* and *Bacillus pulvifaciens* IAM N-19-2 in one and the same fermentation vessel in sequence

Hung. Pat. 150,009

Japan Pat. 303,584

U. S. Pat. 2,993,839

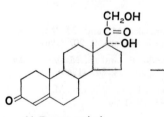

11-Deoxycortisol

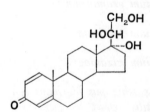

17α, 20β, 21-Trihydroxypregna-1, 4-dien-3-one

Alcaligenes sp.

Corynebacterium simplex

Mycobacterium lacticola ATCC 9626

Pseudomonas oleovorans

Sutter, D., W. Charney, P. L. O'Neill, F. Carvajal, H. L. Herzog and E. B. Hershberg, J. Org. Chem., **22**, 578 (1957)

U. S. Pat. 3,037,915

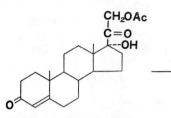

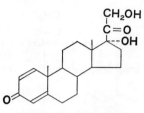

11-Deoxycortisol 21-acetate

17α, 21-Dihydroxypregna-
1, 4-diene-3, 20-dione

Mycobacterium lacticola
Mycobacterium smegmatis

Belg. Pat. 538,327

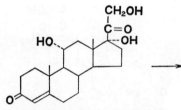

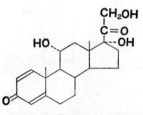

11α, 17α, 21-Trihydroxypregn-
4-ene-3, 20-dione

11α, 17α, 21-Trihydroxypregna-
1, 4-diene-3, 20-dione

Corynebacterium simplex

U. S. Pat. 2,957,893

Fusarium solani, conidia

Vézina, C., S. N. Sehgal and K. Singh, Appl.
Microbiol., **11**, 50 (1963)

Rhizoctonia ferrugena

U. S. Pat. 2,968,595

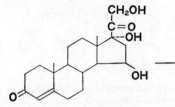

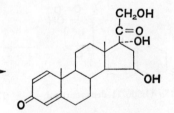

15β-Hydroxy-11-deoxycortisol

15β, 17α, 21-Trihydroxypregna-
1, 4-diene-3, 20-dione

Bacillus sphaericus

U. S. Pat. 2,958,631

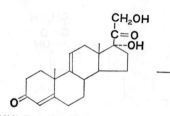

9(11)-Dehydro-11-deoxycortisol

Protaminobacter alboflavum
Protaminobacter rubrum

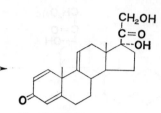

17α, 21-Dihydroxypregna-1,
4, 9(11)-triene-3, 20-dione

U. S. Pat. 2,776,927

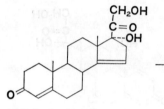

14-Dehydro-11-deoxycortisol

Protaminobacter alboflavum
Protaminobacter rubrum

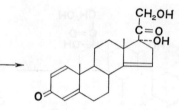

17α, 21-Dihydroxypregna-
1, 4, 14-triene-3, 20-dione

U. S. Pat. 2,776,927

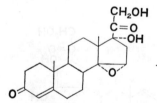

14α, 15α-Oxido-11-deoxycortisol

Protaminobacter alboflavum
Protaminobacter rubrum

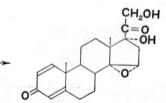

17α, 21-Dihydroxy-14α, 15α-oxido-
pregna-1, 4-diene-3, 20-dione

U. S. Pat. 2,776,927

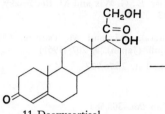

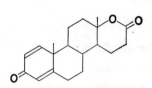

11-Deoxycortisol

1-Dehydrotestololactone

Cylindrocarpon radicicola ATCC 11011
(50%)

Fried, J., R. W. Thoma and A. Klingsberg, J. Am. Chem. Soc., **75**, 5764 (1953)

Pseudomonas chlororaphis IAM 1511

Naito, A., Y. Sato, H. Iizuka and K. Tsuda, Steroids, **3**, 327 (1964)

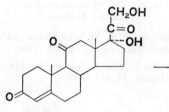

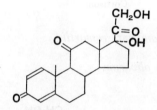

Cortisone

17α, 21-Dihydroxypregna-1, 4-diene-3, 11, 20-trione (Prednisone)

Bacillus cyclooxydans

U. S. Pat. 2,822,318

Bacillus pulvifaciens IAM N-19-2

Iizuka, H., A. Naito and Y. Sato, J. Gen. Appl. Microbiol. (Japan), **7**, 118 (1961)

Bacillus sphaericus

Stoudt, T. H., W. J. McAleer, J. M. Chemerda, M. A. Kozlowski, R. F. Hirschmann, V. Marlatt and R. Miller, Arch. Biochem. Biophys., **59**, 304 (1955)

Corynebacterium simplex ATCC 6946
(70%)

Nobile, A., W. Charney, P. L. Perlman, H. L. Herzog, C. C. Payne, M. E. Tully, M. A. Jevnik and E. B. Hershberg, J. Am. Chem. Soc., **77**, 4184 (1955)

Didymella lycopersici

Vischer, E., Ch. Meystre and A. Wettstein, Helv. Chim. Acta, **38**, 1502 (1955)

Fusarium oxysporum

U. S. Pat. 2,951,016

Fusarium solani

Vischer, E., Ch. Meystre and A. Wettstein, Helv. Chim. Acta, **38**, 835 (1955)

Gliocladium roseum

Shirasaka, M. and M. Tsuruta, Chem. Pharm. Bull. (Japan), **9**, 207 (1961)

Gloeosporium olivarum

Kondo, E. and E. Masuo, J. Agr. Chem. Soc. (Japan), **34**, 847 (1960)

Helminthosporium turcicum

Shirasaka, M. and M. Tsuruta, Chem. Pharm. Bull. (Japan), **9**, 207 (1961)

Mixed culture of *Rhizopus nigricans* and *Bacillus subtilis*

Weisz, E., G. Wix and M. Bodánszky, Naturwiss., **43**, 39 (1956)

Ophiobolus heterostropus

Shirasaka, M. and M. Tsuruta, Chem. Pharm. Bull. (Japan), **9**, 207 (1961)

Protaminobacter alboflavum

U. S. Pat. 2,776,927

Protaminobacter rubrum

Pseudomonas boreopolis

Japan Pat. 303,583

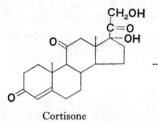

Cortisone

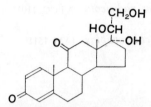

17α, 20β, 21-Trihydroxypregna-1, 4-diene-3, 11-dione

Fusarium solani var. *eumartii*

Szpilfogel, S. A., M. S. DeWinter and W. J. Alsche, Rec. Trav. chim., **75**, 402 (1956)

Gloeosporium olivarum

Kondo, E. and E. Masuo, J. Agr. Chem. Soc. (Japan), **34**, 847 (1960)

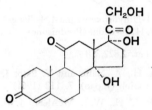

14α-Hydroxycortisone

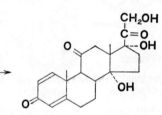

14α, 17α, 21-Trihydroxypregna-1, 4-diene-3, 11, 20-trione

Protaminobacter alboflavum

U. S. Pat. 2,776,927

Protaminobacter rubrum

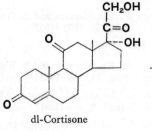

dl-Cortisone

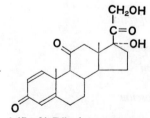

d-17α, 21-Dihydroxypregna-1, 4-diene-3, 11, 20-trione

Didymella lycopersici

Vischer, E., J. Schmidlin and A. Wettstein, Experientia, **12**, 50 (1956)

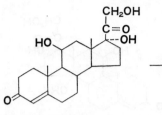 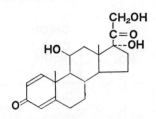

| Cortisol | 11β, 17α, 21-Trihydroxypregna-1, 4-diene-3, 20-dione (Prednisolone) |

Azotomonas fluorescens

U. S. Pat. 2,992,973

Bacillus cyclooxydans

U. S. Pat. 2,822,318

Bacillus pulvifaciens IAM N-19-2 (80%)

Iizuka, H., A. Naito and Y. Sato, J. Gen. Appl. Microbiol. (Japan), **7**, 118 (1961)

Bacillus sphaericus

Stoudt, T. H., W. J. McAleer, J. M. Chemerda, M. A. Kozlowski, R. F. Hirschmann, V. Marlatt and R. Miller, Arch. Biochem. Biophys., **59**, 304 (1955)

Bacterium aromaticus

Japan Pat. 289,323

Corynebacterium equi B-58-1

Japan Pat. 414,359

Corynebacterium simplex ATCC 6946 (Pseudo-crystallofermentation)

Kondo, E. and E. Masuo, J. Gen. Appl. Microbiol. (Japan), **7**, 113 (1961)

Corynebacterium simplex ATCC 6946

Nobile, A., W. Charney, P. L. Perlman, H. L. Herzog, C. C. Payne, M. E. Tully, M. A. Jevnik and E. B. Hershberg, J. Am. Chem. Soc., **77**, 4184 (1955)

Flavobacterium aquatile

Japan Pat. 305,166

Graphiola cylindrica

Kondo, E. and E. Masuo, J. Agr. Chem. Soc. (Japan), **34**, 847 (1960)

Mixed culture of *Rhizopus nigricans* and *Bacillus subtilis*

Weisz, E., G. Wix and M. Bodánszky, Naturwiss, **43**, 39 (1956)

Mycobacterium smegmatis

Brit. Pat. 787,410

Mycobacterium sp.

Japan Pat. 271,310

Nocardia corallina

U. S. Pat. 3,087,864

Nocardia sp.

Peterson, G. E., R. W. Thoma, D. Perlman and J. Fried, J. Bacteriol., **74**, 684 (1957)

Nocardia sp.

Japan Pat. 272,229

Protaminobacter alboflavum

U. S. Pat. 2,776,927

Protaminobacter rubrum

Pseudomonas chlororaphis IAM 1511

Naito, A., Y. Sato, H. Iizuka and K. Tsuda, Steroids, **3**, 327 (1964)

Pseudomonas fluorescens

Brit. Pat. 859,694

Rhizoctonia ferrugena

U. S. Pat. 2,968,595

Streptomyces lavendulae

U. S. Pat. 2,793,164

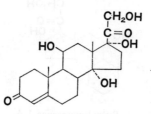

14α-Hydroxycortisol

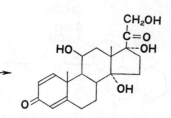

11β, 14α, 17α, 21-Tetrahydroxy-
pregna-1, 4-diene-3, 20-dione

Protaminobacter alboflavum
Protaminobacter rubrum

U. S. Pat. 2,776,927

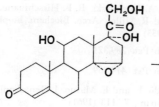

14α, 15α-Oxidocortisol

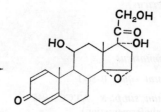

11β, 17α, 21-Trihydroxy-14α, 15α-
oxidopregna-1, 4-diene-3, 20-dione

Protaminobacter alboflavum
Protaminobacter rubrum

U. S. Pat. 2,776,927

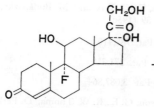

9α-Fluorocortisol

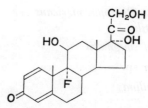

9α-Fluoro-11β, 17α, 21-trihydroxy-
pregna-1, 4-diene-3, 20-dione

Bacillus cyclooxydans
Bacillus sphaericus

U. S. Pat. 2,822,318

Stoudt, T. H., W. J. McAleer, J. M. Chemerda,
M. A. Kozlowski, R. F. Hirschmann, V. Marlatt
and R. Miller, Arch. Biochem. Biophys., **59**, 304
(1955)

Protaminobacter alboflavum
Protaminobacter rubrum

U. S. Pat. 2,776,927

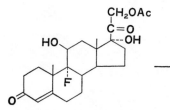

CH₂OAc

9α-Fluorocortisol 21-acetate

Didymella lycopersici

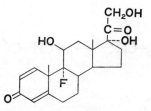

9α-Fluoro-11β, 17α, 21-trihydroxy-
pregna-1, 4-diene-3, 20-dione

Vischer, E., Ch. Meystre and A. Wettstein, Helv.
Chim. Acta, **38**, 1502 (1955)

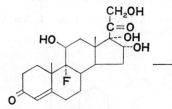

9α-Fluoro-16α-hydroxycortisol

Bacterium havaniensis
Bacterium mycoides
Mycobacterium rhodochrous (65%)

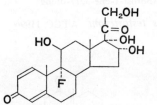

9α-Fluoro-11β, 16α, 17α, 21-tetrahydroxy-
pregna-1, 4-diene-3, 20-dione

U. S. Pat. 3,037,914

U. S. Pat. 3,037,912

Thoma, R. W., J. Fried, S. Bonanno and P.
Grabowich, J. Am. Chem. Soc., **79**, 4818 (1957)

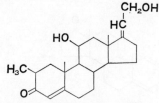

2-Methyl-11β, 21-dihydroxy-
pregna-4, 17(20)-dien-3-one

Septomyxa affinis

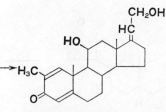

2-Methyl-11β, 21-dihydroxypregna-
1, 4, 17 (20)-trien-3-one

U. S. Pat. 3,009,937

(d) $-CH_2-CH\big< \longrightarrow -CH=C\big<$

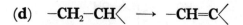

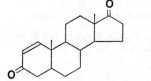

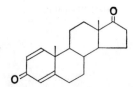

Androst-1-ene-3, 17-dione	Androsta-1, 4-diene-3, 17-dione

Pseudomonas testosteroni ATCC 11996   Levy, H. R. and P. Talalay, J. Am. Chem. Soc., **79**, 2658 (1957)

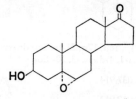

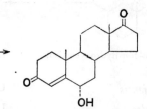

3β-Hydroxy-5α, 6α-oxido-
androstan-17-one

6α-Hydroxyandrost-4-ene-3, 17-dione

Nocardia restrictus No. 545   Lee, S. S. and C. J. Sih, Biochemistry, **3**, 1267 (1964)

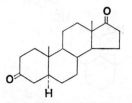

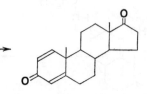

5α-Androstane-3, 17-dione   Androsta-1, 4-diene-3, 17-dione

Mycobacterium smegmatis   Brit. Pat. 850,951

Pseudomonas testosteroni ATCC 11996   Levy, H. R. and P. Talalay, J. Am. Chem. Soc., **79**, 2658 (1957)

C. REDUCTION

(a) $\rangle$C=O $\longrightarrow$ $\rangle$CH–OH

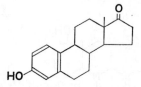

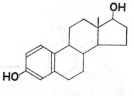

Estrone

α-Estradiol

Baker's yeast (67%)
Press yeast (70%)

Mamoli L., Ber., **71B**, 2696 (1938)
Wettstein A., Helv. Chim. Acta, **22**, 250 (1939)

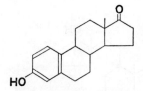

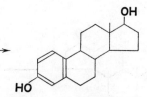

dl-Estrone

d-Estradiol

Press yeast (*Saccharomyces* sp.)

Vischer, E., J. Schmidlin, and A. Wettstein, Experientia, **12**, 50 (1956)

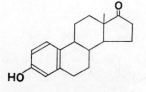

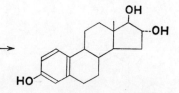

Estrone

16α-Hydroxyestradiol

Streptomyces halstedii ATCC 13499, NRRL B-2138

Streptomyces mediocidicus ATCC 13278

Kita, D. A., J. L. Sardinas and G. M. Shull, Nature, **190**, 627 (1961)

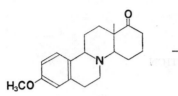

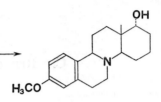

8-Aza-D-homoestrone 3-methylether

8-Aza-D-homoestradiol
3-methylether (17α-OH)

Aspergillus ochraceus

Curtis, P. J., Biochem. J. **97**, 148 (1965)

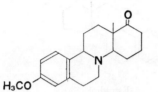

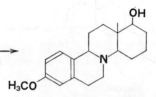

8-Aza-D-homoestrone 3-methylether

8-Aza-D-homoestradiol
3-methylether (17β-OH)

Aspergillus ochraceus

Curtis, P. J., Biochem. J., **97**, 148 (1965)

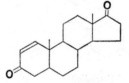

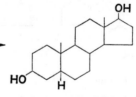

Androst-1-ene-3, 17-dione

5β-Androstane-3β, 17β-diol

Baker's yeast (25%)

Butenandt, A., H. Dannenberg and L. A. Surányi,
Ber., **73B**, 818 (1940)

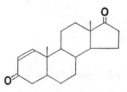

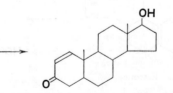

Androst-1-ene-3, 17-dione

17β-Hydroxyandrost-1-en-3-one

Baker's yeast (83%)

Butenandt, A. and H. Dannenberg, Ber., **71B**, 1681
(1938)

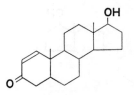

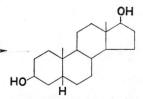

17β-Hydroxyandrost-1-en-3-one   5β-Androstane-3β, 17β-diol

Baker's yeast   Butenandt, A., H. Dannenberg and L. A. Surányi, Ber, **73B**, 818 (1940)

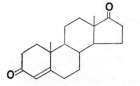

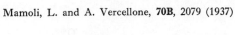

Androst-4-ene-3, 17-dione   Testosterone

Yeast   Mamoli, L. and A. Vercellone, Ber. **70B**, 470 (1937)

Top yeast   Mamoli, L. and A. Vercellone, **70B**, 2079 (1937)

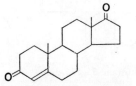

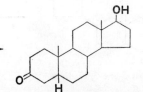

Androst-4-ene-3, 17-dione   17β-Hydroxy-5β-Androstan-3-one

Clostridium   lentoputrescens   (Bacillus putrificus) (59%)   Mamoli, L., R. Koch and H. Teschen, Z. Physiol. Chem., **261**, 287 (1939)

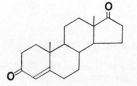

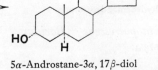

Androst-4-ene-3, 17-dione   5α-Androstane-3α, 17β-diol

Clostridium   lentoputrescens   (Bacillus putrificus) (13%)   Mamoli, L., R. Koch and H. Teschen, Z. Physiol. Chem., **261**, 287 (1939)

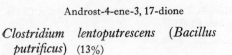

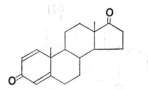

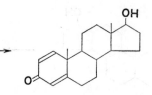

Androsta-1, 4-diene-3, 17-dione

17β-Hydroxyandrosta-1, 4-dien-3-one

Fusarium lateritium

Čapek, A., O. Hanč and M. Tadra, Folia Micro-biol., **8**, 120 (1963)

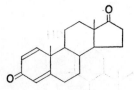

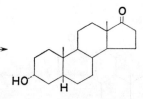

Androsta-1, 4-diene-3, 17-dione

3α-Hydroxy-5β-androstan-17-one

Clostridium paraputrificum

Schubert, K., J. Schlegel and C. Hörhold, Z. Physiol. Chem., **332**, 310 (1963)

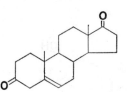

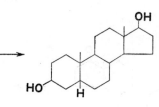

Androst-5-ene-3, 17-dione

5β-Androstane-3β, 17β-diol

Top yeast

Mamoli, L. and A. Vercellone, Ber., **70B**, 2079 (1937)

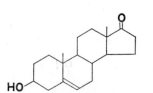

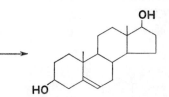

Dehydroepiandrosterone

Androst-5-ene-3β, 17β-diol

Clostridium lentoputrescens (*Bacillus putrificus*) (67%)

Mamoli, L., R. Koch and H. Teschen, Z. Physiol. Chem., **261**, 287 (1939)

Yeast (18%)

Mamoli, L. and A. Vercellone, Z. Physiol. Chem., **245**, 93 (1937)

Yeast

U. S. Pat. 2,186,906

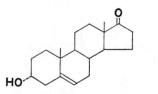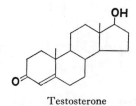

3α-Hydroxyandrost-5-en-17-one   Testosterone

Baker's yeast (81%)   Mamoli, L., Ber., **71B**, 2278 (1938)

Impoverished yeast   U. S. Pat. 2,236,574

Oxidizing bacteria   Mamoli, L., Ber., **71B**, 2278 (1938)

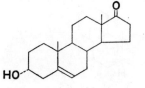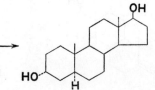

3α-Hydroxyandrost-5-en-17-one   5α-Androstane-3α, 17β-diol

Bacteria   Schramm, G. and L. Mamoli, Ber., **71B**, 1322 (1938)

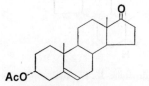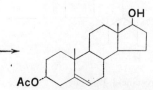

3α-Hydroxyandrost-5-en-   3α, 17β-Dihydroxyandrost-
17-one 3-acetate   5-ene 3-acetate

Yeast   Mamoli, L. Ber., **71B**, 2696 (1938)

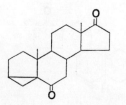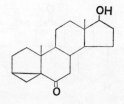

3, 5-Cycloandrostane-6, 17-dione   3, 5-Cyclo-17β-hydroxyandrostan-6-one

Yeast (50%)   Butenandt, A. and L. A. Surányi, Ber., **75B**, 591
                          (1942)

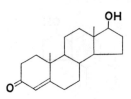
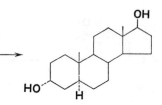

Testosterone   5α-Androstane-3α, 17β-diol

Clostridium lentoputrescens (*Bacillus putrificus*)

Mamoli, L., R. Koch and H. Teschen, Z. Physiol. Chem., **261**, 287 (1939)

Putrefactive bacteria (60%)

Mamoli, L. and G. Schramm, Ber., **71B**, 2083 (1938)

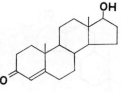
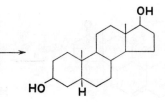

Testosterone   5β-Androstane-3β, 17β-diol

Bacteria (14%)

Mamoli, L. and G. Schramm, Ber., **71B**, 2698 (1938)

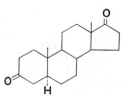
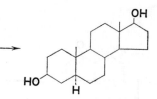

5α-Androstane-3, 17-dione   5α-Androstane-3β, 17β-diol

Clostridium lentoputrescens (*Bacillus putrificus*) (35.5%)

Mamoli, L., R. Koch and H. Teschen, Z. Physiol. Chem., **261**, 287 (1939)

Top yeast

Mamoli, L. and A. Vercellone, Z. Physiol. Chem., **245**, 93 (1937)

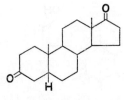
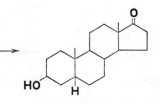

5β-Androstane-3, 17-dione   3α-Hydroxy-5β-androstan-17-one

Clostridium lentoputrescens (*Bacillus putrificus*) (36.4%)

Mamoli, L., R. Koch and H. Teschen, Z. Physiol. Chem., **261**, 287 (1939)

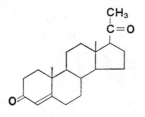 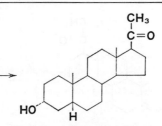

Progesterone   3α-Hydroxy-5β-pregnan-20-one

Alternaria bataticola   Shirasaka, M. and M. Ozaki, J. Agr. Chem. Soc.,
                                       (Japan), **35**, 200 (1961)

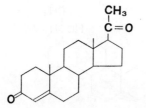 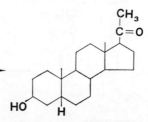

Progesterone   3β-Hydroxy-5β-pregnan-20-one

Alternaria bataticola   Shirasaka, M. and M. Ozaki, J. Agr. Chem. Soc.,
                                       (Japan), **35**, 200 (1961)

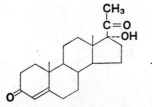 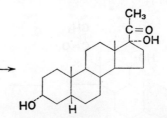

17α-Hydroxyprogesterone   3α, 17α-Dihydroxy-5β-pregnan-20-one

Alternaria bataticola   Shirasaka, M. and M. Ozaki, J. Agr. Chem. Soc.,
                                       (Japan), **35**, 200 (1961)

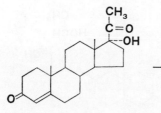 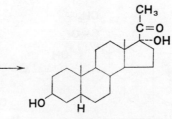

17α-Hydroxyprogesterone   3β, 17α-Dihydroxy-5β-pregnan-20-one

Alternaria bataticola   Shirasaka, M. and M. Ozaki, J. Agr. Chem. Soc.,
                                       (Japan), **35**, 200 (1961)

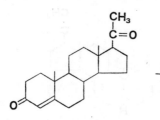

Progesterone

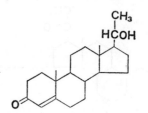

20α-Hydroxypregn-4-en-3-one

Rhodotorula longissima strain Schering OFU No. 2

Chang, V. M. and D. R. Idler, Can. J. Biochem. Physiol., **39**, 1277 (1961)

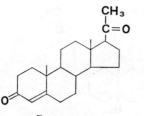

Progesterone

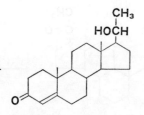

20β-Hydroxypregn-4-en-3-one

Penicillium lilacinum

Sebek, O. K., L. M. Reineke and D. H. Peterson, J. Bacteriol., **83**, 1327 (1962)

Streptomyces lavendulae strain Rutgers Univ. No. 3440-14

Fried, J., R. W. Thoma and A. Klingsberg, J. Am. Chem. Soc., **75**, 5764 (1953)

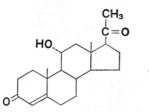

11α-Hydroxyprogesterone

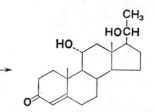

11α, 20β-Dihydroxypregn-4-en-3-one

Penicillium lilacinum

Sebek, O. K., L. M. Reineke and D. H. Peterson, J. Bacteriol., **83**, 1327 (1962)

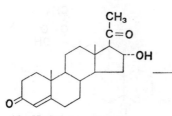

16α-Hydroxyprogesterone

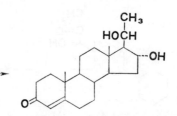

16α, 20β-Dihydroxypregn-4-en-3-one

Streptomyces lavendulae

Fried, J., R. W. Thoma, D. Perlman, J. E. Herz and A. Borman, Recent Progr. Hormone Res., **11**, 149 (1955)

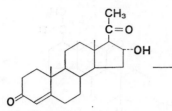

16α-Hydroxyprogesterone

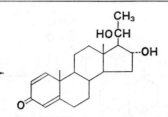

16α, 20β-Dihydroxypregna-
1, 4-dien-3-one

Streptomyces lavendulae

Fried, J., R. W. Thoma, D. Perlman, J. E. Herz
and A. Borman, Recent Progr. Hormone Res.,
11, 149 (1955)

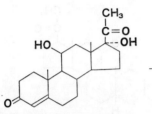

11β, 17α-Dihydroxyprogesterone

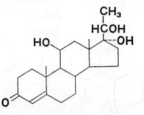

11β, 17α, 20α-Trihydroxypregn-
4-en-3-one

Rhodotorula glutinis IFO 0395 (45%)

Takahashi, T. and Y. Uchibori, Agr. Biol. Chem.
(Japan), **26**, 89 (1962)

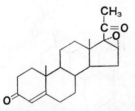

16α, 17α-Oxidoprogesterone

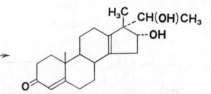

17β-Methyl-16α, 20α-dihydroxy-
18-nor-17α-pregna-4, 13(14)-
dien-3-one

Yeast (60%)

Camerino, B. and R. Modelli, Gazz. Chim. Ital.,
86, 1219 (1956)

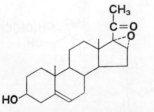

16α, 17α-Oxidopregnenolone

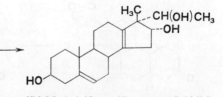

17β-Methyl-18-nor-17α-pregna-5, 13(14)-
diene-3β, 16α, 20α-triol

Yeast (20%)

Camerino, B., R. Modelli and C. Spalla, Gazz.
Chim. Ital., **86**, 1226 (1956)

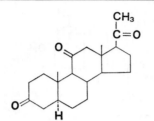

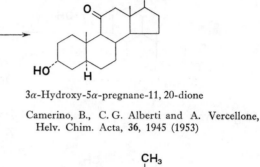

5α-Pregnane-3, 11, 20-trione

3α-Hydroxy-5α-pregnane-11, 20-dione

Yeast (44%)

Camerino, B., C. G. Alberti and A. Vercellone, Helv. Chim. Acta, **36**, 1945 (1953)

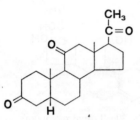

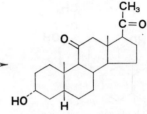

5β-Pregnane-3, 11, 20-trione

3α-Hydroxy-5β-pregnane-11, 20-dione

Yeast (60%)

Camerino, B., C. G. Alberti and A. Vercellone, Helv. Chim. Acta, **36**, 1945 (1953)

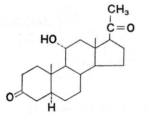

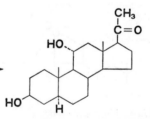

11α-Hydroxy-5α-pregnane-3, 20-dione

3β, 11α-Dihydroxy-5α-pregnan-20-one

Yeast (60%)

Camerino, B., C. G. Alberti and A. Vercellone, Helv. Chim. Acta, **36**, 1945 (1953)

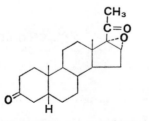

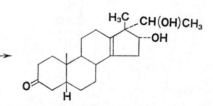

16α, 17α-Oxido-5β-pregnane-3, 20-dione

17β-Methyl-16α, 20α-dihydroxy-18-nor-5β-pregn-13(14)-en-3-one

Yeast

Camerino, B. and A. Vercellone, Gazz. Chim. Ital., **86**, 260 (1956)

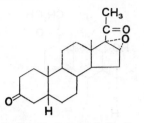 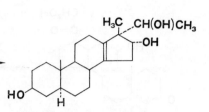

16α, 17α-Oxido-5β-pregnane-3, 20-dione

17β-Methyl-18-nor-5α-pegn-13(14)-ene-3β, 16α, 20α-triol

Yeast

Camerino, B. and A. Vercellone, Gazz. Chim. Ital., **86**, 260 (1956)

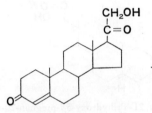 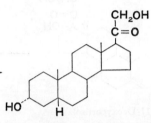

11-Deoxycorticosterone

3α, 21-Dihydroxy-5β-pregnan-20-one

Alternaria bataticola

Shirasaka, M. and M. Ozaki, J. Agr. Chem. Soc. (Japan), **35**, 200 (1961)

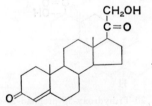 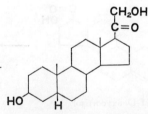

11-Deoxycorticosterone

3β, 21-Dihydroxy-5β-pregnan-20-one

Alternaria bataticola

Shirasaka, M. and M. Ozaki, J. Agr. Chem. Soc. (Japan), **35**, 200 (1961)

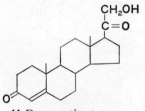 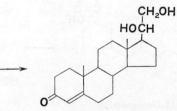

11-Deoxycorticosterone

20β, 21-Dihydroxypregn-4-en-3-one

Streptomyces sp.

Peterson, D. H., Perspectives and Horizons in Microbiology p121 (1955)

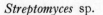

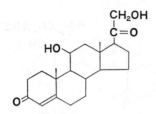

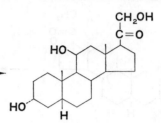

Corticosterone

3α, 11β, 21-Trihydroxy-5β-pregnan-20-one

Alternaria bataticola

Shirasaka, M. and M. Ozaki, J. Agr. Chem. Soc. (Japan), **35**, 200 (1961)

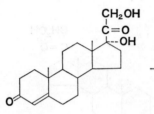

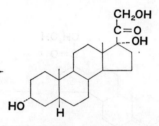

11-Deoxycortisol

3α, 17α, 21-Trihydroxy-5β-pregnan-20-one

Alternaria bataticola

Shirasaka, M. and M. Ozaki, J. Agr. Chem. Soc. (Japan), **35**, 200 (1961)

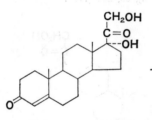

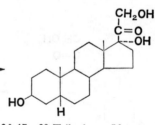

11-Deoxycortisol

3β, 17α, 20-Trihydroxy-5β-pregnan-20-one

Alternaria bataticola

Shirasaka, M. and M. Ozaki, J. Agr. Chem. Soc. (Japan), **35**, 200 (1961)

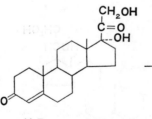

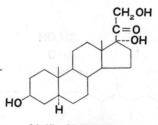

11-Deoxycortisol

3β, 17α, 21-Trihydroxy-5α-pregnan-20-one

Streptomyces aureus ATCC 3309

Kondo, E., T. Mitsugi and E. Masuo, Agr. Biol. Chem. (Japan), **26**, 22 (1962)

— 150 —

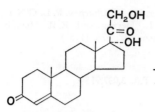

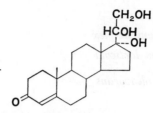

11-Deoxycortisol

17α, 20α, 21-Trihydroxypregn-
4-en-3-one

Rhodotorula glutinis   IFO 0395 (65%)

Takahashi, T. and Y. Uchibori, Agr. Biol. Chem.
(Japan), **26**, 89 (1962)

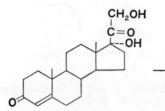

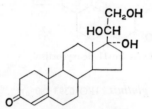

11-Deoxycortisol

17α, 20β, 21-Trihydroxypregn-
4-en-3-one

Candida pulcherrima   IFO 0964

Takahashi, T. and Y. Uchibori, Agr. Biol. Chem.
(Japan), **26**, 89 (1962)

Curvularia lunata   NRRL 2380

Townsley, J. D., H. J. Brodie, M. Hayano and R. I.
Dorfman, Steroids, **3**, 341 (1964)

Didymella lycopersici   ATCC 11847 (5%)

Sehgal, S. N., K. Singh and C. Vézina, Steroids,
2, 93 (1963)

Didymella lycopersici,   conidia

Vézina, C., S. N. Sehgal and K. Singh, Appl.
Microbiol., **11**, 50 (1963)

Epicoccum oryzae

Shull, G. M., Trans. N. Y. Acad. Sci., **19**, 147 (1956)

Pseudomonas fluorescens

Brit. Pat. 859,694

Pythium ultimum

Shirasaka, M. and M. Ozaki, J. Agr. Chem. Soc.
(Japan), **35**, 206 (1961)

Sporotrichum gougeroti   IFO 5982 (60∼
70%)

Takahashi, T. and Y. Uchibori, Agr. Biol. Chem.,
(Japan), **26**, 89 (1962)

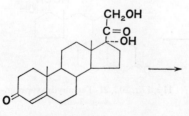

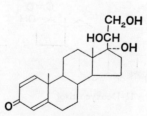

11-Deoxycortisol

17α, 20β, 21-Trihydroxypregna-
1, 4-dien-3-one

Alcaligenes sp.

Corynebacterium simplex

Mycobacterium lacticola

Pseudomonas oleovorans

Sutter, D., W. Charney, P. L. O'Neill, F. Carvajal, H. L. Herzog and E. B. Hershberg, J. Org. Chem., **22**, 578, (1957)

U. S. Pat. 3,037,915

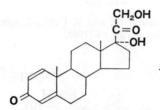

1-Dehydro-11-deoxycortisol

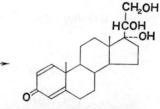

17α, 20α, 21-Trihydroxypregna-
1, 4-dien-3-one

Rhodotorula glutinis IFO 0395 (80%)

Takahashi, T. and Y. Uchibori, Agr. Biol. Chem. (Japan), **26**, 89 (1962)

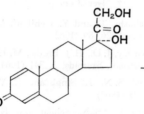

1-Dehydro-11-deoxycortisol

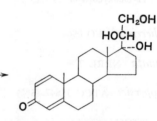

17α, 20β, 21-Trihydroxypregna-
1, 4-dien-3-one

Candida pulcherrima IFO 0964 (40~80%)
Sporotrichum gougeroti IFO 5982 (40~
80%)

Takahashi, T. and Y. Uchibori, Agr. Biol. Chem. (Japan), **26**, 89 (1962)

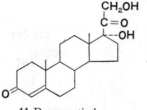

11-Deoxycortisol

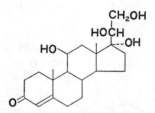

11β, 17α, 20β, 21-Tetrahydroxypregn-
4-en-3-one

Curvularia lunata

Shull, G. M., Trans, N. Y. Acad. Sci., **19**, 147 (1956)

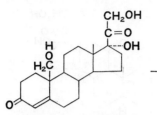

19-Hydroxy-11-deoxycortisol

Rhodotorula glutinis IFO 0395 (10%)

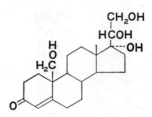

17α, 19, 20α, 21-Tetrahydroxypregn-
4-en-3-one

Takahashi, T. and Y. Uchibori, Agr. Biol. Chem.
(Japan), **26**, 89 (1962)

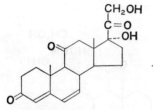

17α, 21-Dihydroxypregna-4, 6-
diene-3, 11, 20-trione

Curvularia lunata NRRL 2380 (20~25%)

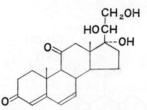

17α, 20β, 21-Trihydroxypregna-
4, 6-diene-3, 11-dione

Gould, D., J. Ilavsky, R. Gutekunst and E. B.
Hershberg, J. Org. Chem., **22**, 829 (1957)

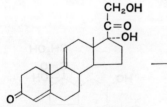

17α, 21-Dihydroxypregna-4, 9(11)-
diene-3, 20-dione

Curvularia lunata

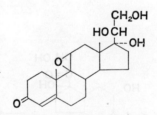

17α, 20β, 21-Trihydroxy-9β, 11β-
oxidopregn-4-en-3-one

Shull, G. M., Trans. N. Y. Acad. Sci., **19**, 147 (1956)

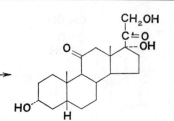

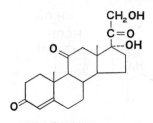

Cortisone

3α, 17α, 21-Trihydroxy-5β-pregnane-
11, 20-dione

Alternaria bataticola

Shirasaka, M. and M. Ozaki, J. Agr. Chem. Soc.
(Japan), **35**, 200 (1961)

Catenabacterium catenaforme

Prévot, A.-R., M.-M. Janot and N. D. Tam,
Compt. Rend., Soc. Biol., **256**, 3785 (1963)

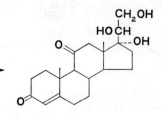

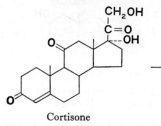

Cortisone

17α, 20β, 21-Trihydroxypregn-
4-ene-3, 11-dione

Fusarium solani var. *eumartii*

Szpilfogel, S. A., M. S. DeWinter and W. J. Alsche,
Rec. Trav. chim., **75**, 402 (1956)

Gloeosporium olivarum

Kondo, E. and E. Masuo, J. Agr. Chem. Soc.
(Japan), **34**, 847 (1960)

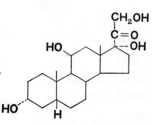

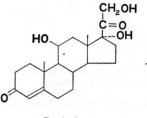

Cortisol

3α, 11β, 17α, 21-Tetrahydroxy-
5β-pregnan-20-one

Alternaria bataticola

Shirasaka, M. and M. Ozaki, J. Agr. Chem. Soc.
(Japan), **35**, 200 (1961)

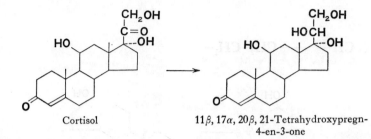

Cortisol

11β, 17α, 20β, 21-Tetrahydroxypregn-
4-en-3-one

Streptomyces hydrogenans

Schmidt-Thomé, J., Angew. Chem., **69**, 238 (1957)

(b) $-CH=CH- \longrightarrow -CH_2-CH_2-$

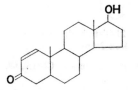 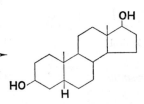

17β-Hydroxyandrost-1-en-3-one   5β-Androstane-3β, 17β-diol

Baker's yeast

Butenandt, A., H. Dannenberg and L. A. Surányi, Ber., **73 B**, 818 (1940)

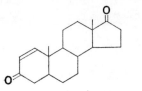 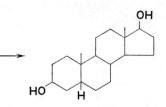

Androst-1-ene-3, 17-dione   5β-Androstane-3β, 17β-diol

Baker's yeast

Butenandt, A., H. Dannenberg and L. A. Surányi, Ber., **73 B**, 818 (1940)

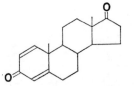 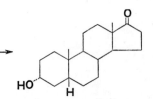

Androsta-1, 4-diene-3, 17-dione   3α-Hydroxy-5β-androstan-17-one

Clostridium paraputrificum

Schubert, K., J. Schlegel and C. Hörhold, Z. physiol. Chem., **332**, 310 (1963)

(c) $-CH=C\!\!\big\langle \longrightarrow -CH_2-CH\!\!\big\langle$

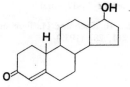

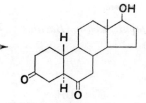

19-Nortestosterone

17β-Hydroxy-19-nor-5α-androstane-
3, 6-dione

Rhizopus reflexus

U. S. 2,692,273

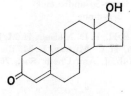

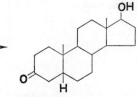

Testosterone

17β-Hydroxy-5β-androstan-3-one

Clostridium lentoputrescens (Bacillus putrificus) (70~80%)

Mamoli, L., R. Koch and H. Teschen, Z. Physiol. Chem., **261**, 287 (1939)

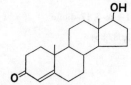

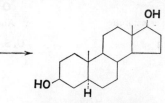

Testosterone

5α-Androstane-3β, 17β-diol

Putrefactive bacteria

Mamoli, L. and G. Schramm, Ber., **71B**, 2698 (1938)

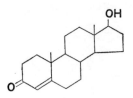

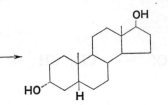

Testosterone

5β-Androstane-3α-17β-diol

Clostridium lentoputrescens (Bacillus putrificus) (10%)

Mamoli, L., R. Koch and H. Teschen, Z. Physiol. Chem., **261**, 287 (1939)

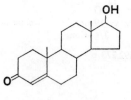

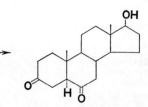

Testosterone

17β-Hydroxy-5α-androstane-3, 6-dione

Rhizopus reflexus

Eppstein, S. H., P. D. Meister, H. M. Leigh, D. H. Peterson, H. C. Murray, L. M. Reineke and A. Weintraub, J. Am. Chem. Soc., **76**, 3174 (1954)

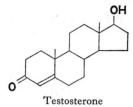

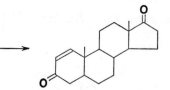

Testosterone

Androst-1-ene-3, 17-dione

Nocardia corallina

U. S. Pat. 3,087,864

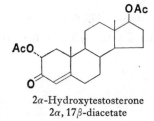

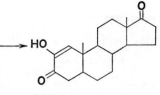

2α-Hydroxytestosterone 2α, 17β-diacetate

2-Hydroxyandrost-1-ene-3, 17-dione

Nocardia corallina

U. S. Pat. 3,087,864

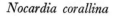

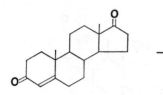

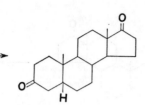

Androst-4-ene-3, 17-dione

5β-Androstane-3, 17-dione

Clostridium lentoputrescens (*Bacillus putrificus*) (70～80%)

Mamoli, L., R. Koch and H. Teschen, Z. Physiol. Chem., **261**, 287 (1939)

Putrefactive bacteria

Mamoli, L. and G. Schramm, Ber., **71 B**, 2083 (1938)

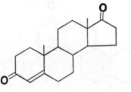

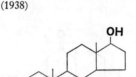

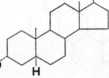

Androst-4-ene-3, 17-dione

5β-Androstane-3α, 17β-diol

Clostridium lentoputrescens (*Bacillus putrificus*) (13%)

Mamoli, L., R. Koch and H. Teschen, Z. Physiol. Chem., **261**, 287 (1939)

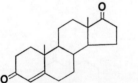

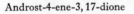

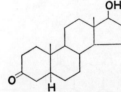

Androst-4-ene-3, 17-dione

17β-Hydroxy-5β-androstan-3-one

Clostridium lentoputrescens (*Bacillus putrificus*) (39%)

Mamoli, L., R. Koch and H. Teschen, Z. Physiol. Chem., **261**, 287 (1939)

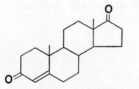

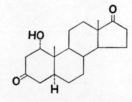

Androst-4-ene-3, 17-dione

1α-Hydroxy-5α-androstane-3, 17-dione

Penicillium sp. ATCC 12556

Dodson, R. M., A. H. Goldkamp and R. D. Muir, J. Am. Chem. Soc., **82**, 4026 (1960)

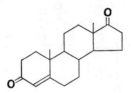

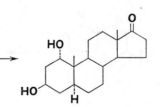

Androst-4-ene-3, 17-dione

1α, 3β-Dihydroxy-5α-androstan-17-one

Penicillium sp. ATCC 12556

Dodson, R. M., A. H. Goldkamp and R. D. Muir, J. Am. Chem. Soc., **82**, 4026 (1960)

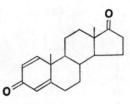

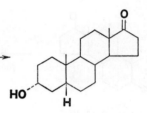

Androsta-1, 4-diene-3, 17-dione

3α-Hydroxy-5β-androstan-17-one

Clostridium paraputrificum

Schubert, K., J. Schlegel and C. Hörhold, Z. Physiol. Chem., **332**, 310 (1963)

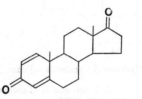

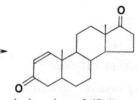

Androsta-1, 4-diene-3, 17-dione

Androst-1-ene-3, 17-dione

Clostridium paraputrificum

Schubert, K., J. Schlegel and C. Hörhold, Z. Physiol. Chem., **332**, 310 (1963)

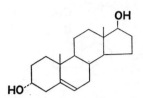
Androst-5-ene-3α, 17β-diol

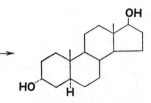
5α-Androstane-3α, 17β-diol

Bacteria

Schramm, G. and L. Mamoli, Ber., **71 B**, 1322 (1938)

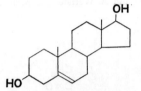
Androst-5-ene-3β, 17β-diol

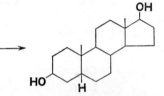
5β-Androstane-3β, 17β-diol

Yeast

U. S. Pat. 2,186,906

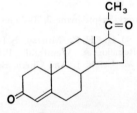
Progesterone

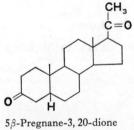
5β-Pregnane-3, 20-dione

Clostridium lentoputrescens (*Bacillus putrificus*) (86%)

Mamoli, L., R. Koch and H. Teschen, Z. Physiol. Chem., 261, 287 (1939)

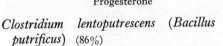

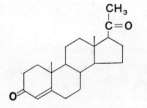
Progesterone

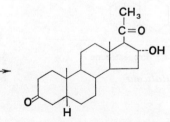
16α-Hydroxy-5β-pregnane-3, 20-dione

Unidentified Actinomycetes

Perlman, D., E. Titus and J. Fried, J. Am. Chem. Soc., **74**, 2126 (1952)

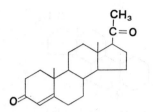

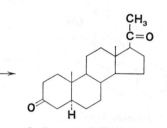

Progesterone

5α-Pregnane-3, 20-dione

Cortinarius evernius C-351

Schuytema, E. C., M. P. Hargie, D. J. Siehr, I. Merits, J. R. Schenck, M. S. Smith and E. L. Varner, Appl. Microbiol., **11**, 256 (1963)

Streptomyces griseus

Vischer, E. and A. Wettstein, Experientia, **16**, 355 (1960)

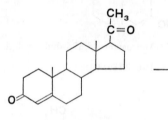

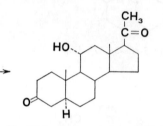

Progesterone

11α-Hydroxy-5α-pregnane-3, 20-dione

Rhizopus nigricans ATCC 6227b (4.0%)

Peterson, D. H., H. C. Murray, S. H. Eppstein, L. M. Reineke, A. Weintraub, P. D. Meister and H. M. Leigh, J. Am. Chem. Soc., **74**, 5933 (1952)

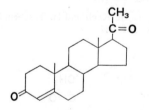

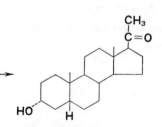

Progesterone

3α-Hydroxy-5β-pregnan-20-one

Alternaria bataticola

Shirasaka, M. and M. Ozaki, J. Agr. Chem. Soc. (Japan), **35**, 200 (1961)

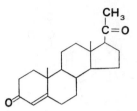

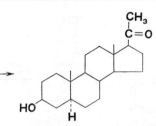

Progesterone

3β-Hydroxy-5α-pregnan-20-one

Streptomyces griseus

Vischer, E. and A. Wettstein, Experientia, **16**, 355 (1960)

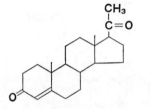

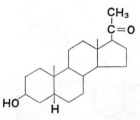

Progesterone

3β-Hydroxy-5β-pregnan-20-one

Alternaria bataticola

Shirasaka, M. and M. Ozaki, J. Agr. Chem. Soc. (Japan), **35** 200 (1961)

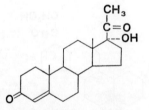

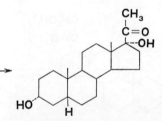

17α-Hydroxyprogesterone

3α, 17α-Dihydroxy-5β-pregnan-20-one

Alternaria bataticola

Shirasaka, M. and M. Ozaki, J. Agr. Chem. Soc. (Japan), **35**, 200 (1961)

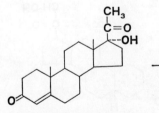

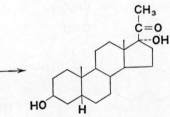

17α-Hydroxyprogesterone

3β, 17α-Dihydroxy-5β-pregnan-20-one

Alternaria bataticola

Shirasaka, M. and M. Ozaki, J. Agr. Chem. Soc. (Japan), **35**, 200 (1961)

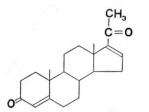

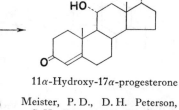

16-dehydroprogesterone

11α-Hydroxy-17α-progesterone

Rhizopus nigricans   ATCC 6227b (24%)

Meister, P. D., D. H. Peterson, H. C. Murray, S. H. Eppstein, L. M. Reineke, A. Weintraub and H. M. Leigh, J. Am. Chem. Soc., **75**, 55 (1953)

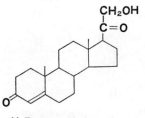

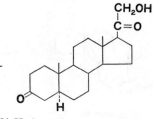

11-Deoxycorticosterone

21-Hydroxy-5α-pregnane-3, 20-dione

Streptomyces griseus   (40%)

Vischer, E. and A. Wettstein, Experientia, **16**, 355 (1960)

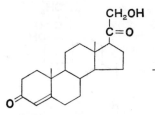

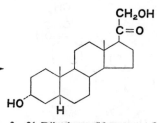

11-Deoxycorticosterone

3α, 21-Dihydroxy-5β-pregnan-20-one

Alternaria bataticola

Shirasaka, M. and M. Ozaki, J. Agr. Chem. Soc. (Japan), **35**, 200 (1961)

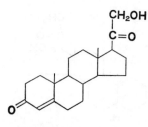

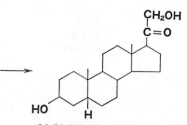

11-Deoxycorticosterone

3β, 21-Dihydroxy-5β-pregnan-20-one

Alternaria bataticola

Shirasaka, M. and M. Ozaki, J. Agr. Chem. Soc. (Japan), **35**, 200 (1961)

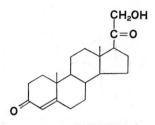

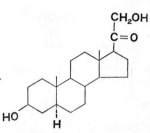

11-Deoxycorticosterone

3β, 21-Dihydroxy-5α-pregnan-20-one

Streptomyces griseus (85%)

Vischer, E. and A. Wettstein, Experientia, **16**, 355 (1960)

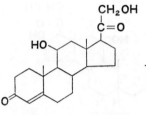

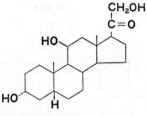

Corticosterone

3α, 11β, 21-Trihydroxy-
5β-pregnan-20-one

Alternaria bataticola

Shirasaka, M. and M. Ozaki, J. Agr. Chem. Soc. (Japan), **35**, 200 (1961)

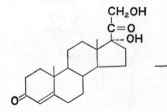

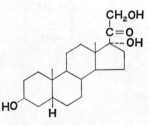

11-Deoxycortisol

3α, 17α, 21-Trihydroxy-
5β-pregnan-20-one

Alternaria bataticola

Shirasaka, M. and M. Ozaki, J. Agr. Chem. Soc. (Japan), **35**, 200 (1961)

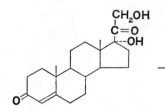

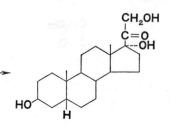

11-Deoxycortisol

3β, 17α, 21-Trihydroxy-
5β-pregnan-20-one

Alternaria bataticola

Shirasaka, M. and M. Ozaki, J. Agr. Chem. Soc.
(Japan), **35**, 200 (1961)

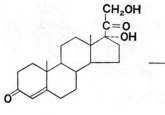

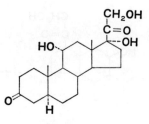

11-Deoxycortisol

11α, 17α, 21-Trihydroxy-5α-
pregnane-3, 20-dione

Rhizopus nigricans (8.2%)

U. S. Pat. 2,602,769

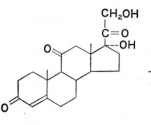

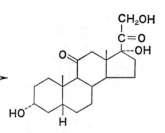

Cortisone

3α, 17α, 21-Trihydroxy-5β-pregnane-
11, 20-dione

Alternaria bataticola

Shirasaka, M. and M. Ozaki, J. Agr. Chem. Soc.
(Japan), **35**, 200 (1961)

Catenabacterium catenaforme

Prévot, A.-R., M.-M. Janot and N.-D. Tam,
Compt. Rend. Soc. Biol., **256**, 3785 (1963)

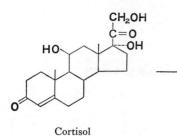

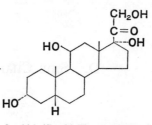

Cortisol

3α, 11β, 17α, 21-Tetrahydroxy-5β-
pregnan-20-one

Alternaria bataticola

Shirasaka, M. and M. Ozaki, J. Agr. Chem. Soc.
(Japan), **35**, 200 (1961)

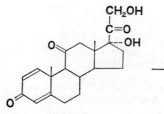

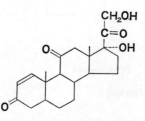

Prednisone

17α, 21-Dihydroxypregn-1-ene-
3, 11, 20-trione

Streptomyces sp. W-3808

Greenspan, G., C. P. Schaffner, W. Charney,
M. J. Gentles and H. L., Herzog, J. Org. Chem.,
26, 1676 (1961)

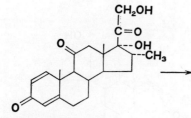

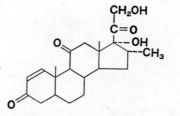

16α-Methylprednisone

16α-Methyl-17α, 21-dihydroxy-
pregn-1-ene-3, 11, 20-trione

Streptomyces sp. W-3808

Greenspan, G., C. P. Schaffner, W. Charney,
M. J. Gentles and H. L. Herzog, J. Org. Chem.,
26, 1676 (1961)

D. SIDE CHAIN DEGRADATION

(a) 17-Alcohol formation

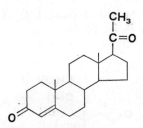

Progesterone

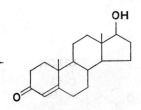

Testosterone

Cladosporium resinae

Fonken, G. S., H. C. Murray and L. M. Reineke, J. Am. Chem. Soc., **82**, 5507 (1960)

Penicillium citrinum (20%)
Penicillium decumbens (60%)

Hanč, O., A. Čapek, M. Tadra, K. Macek and A. Šimek, Arzneimittel-Forsch., **7**, 175 (1957)

Penicillium lilacinum

Sebek, O. K., L. M. Reineke and D. H. Peterson, J. Bacteriol., **83**, 1327 (1962)

Penicillium notatum (62%)

Hanč, O., A. Čapek, M. Tadra, K. Macek and A. Šimek, Arzneimittel-Forsch., **7**, 175 (1957)

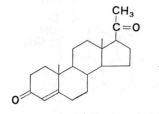

Progesterone

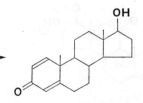

17β-Hydroxyandrosta-1, 4-
dien-3-one

Cylindrocarpon radicicola ATCC 11011

Fried, J., R. W. Thoma and A. Klingsberg, J. Am. Chem. Soc., **75**, 5764 (1953)

Fusarium solani

Nishikawa, M., S. Noguchi and T. Hasegawa, Pharm. Bull. (Japan), **3**, 322 (1955)

Streptomyces lavendulae strain Rutgers
Univ. No. 3440-14 (12%)

Fried, J., R. W. Thoma and A. Klingsberg, J. Am. Chem. Soc., **75**, 5764 (1953)

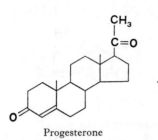

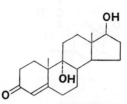

Progesterone

9α-Hydroxytestosterone

Nocardia corallina

Brit. Pat. 862,701

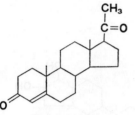

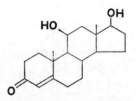

Progesterone

11β-Hydroxytestosterone

Aspergillus tamarii (14%)

Brannon, D. R., J. Martin, A. C. Oehlschlager, N. N. Durham and L. H. Zalkow, J. Org. Chem. **30**, 760 (1965)

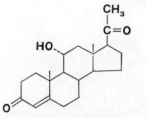

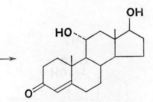

11α-Hydroxyprogesterone

11α-Hydroxytestosterone

Aspergillus chevalieri
Aspergillus oryzae (45%)
Penicillium citrinum

Čapek, A., O. Hanč, K. Macek, M. Tadra and E. Riedl-Tůmová, Naturwiss, **43**, 471 (1956)

Penicillium citrinum
Penicillium decumbens

Hanč, O., A. Čapek, M. Tadra, K. Macek and A. Šimek, Arzneimittel-Forsch., **7**, 175 (1957)

Penicillium lilacinum

Sebek, O. K., L. M. Reineke and D. H. Peterson, J. Bacteriol., **83**, 1327 (1962)

Penicillium notatum

Hanč, O., A. Čapek, M. Tadra, K. Macek and A. Šimek, Arzneimittel-Forsch., **7**, 175 (1957)

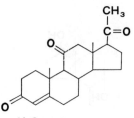
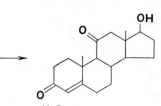

11-Oxoprogesterone　　　　　　　11-Oxotestosterone

Penicillium citrinum　　　　　　Hanč, O., A. Čapek, M. Tadra, K. Macek and
Penicillium decumbens　　　　　A. Šimek, Arzneimittel-Forsch., **7**, 175 (1957)
Penicillium notatum (51%)

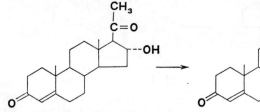
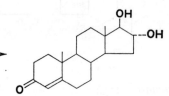

16α-Hydroxyprogesterone　　　　16α-Hydroxytestosterone

Streptomyces lavendulae　　　　Fried, J. R. W. Thoma, D. Perlman, J. E. Herz
and A. Borman, Rec. Progr. Hormone Res., **11**,
149 (1955)

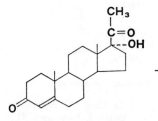
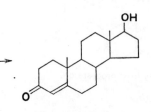

17α-Hydroxyprogesterone　　　　　Testosterone

Penicillium citrinum　　　　　　Hanč, O., A. Čapek, M. Tadra, K. Macek and
Penicillium decumbens　　　　　A. Šimek, Arzneimittel-Forsch., **7**, 175 (1957)
Penicillium notatum

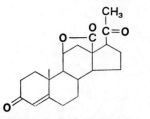 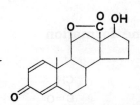

11β-Hydroxy-3, 20-dioxopregn-
4-en-18-oic acid-18, 11-lactone

Fusarium solani

11β, 17β-Dihydroxy-3-oxoandrosta-
1, 4-dien-18-oic acid-18, 11-lactone

Urech, J., E. Vischer and A. Wettstein, Paper,
Meeting Swiss Chem. Soc., September (1961)

(b) 17-Ketone formation

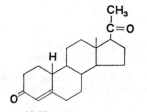

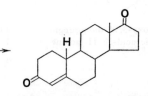

19-Norprogesterone

19-Norandrost-4-ene-3, 17-dione

Streptomyces lavendulae ATCC 8664

Gaul, C., R. I. Dorfman and S. R. Stitch, Biochem.. Biophys. Acta., **49**, 387 (1961)

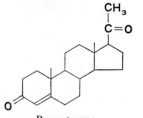

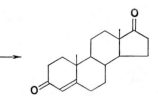

Progesterone

Androst-4-ene-3, 17-dione

Aspergillus flavus

Peterson, D. H., S. H. Eppstein, P. D. Meister, H. C. Murray, H. M. Leigh, A. Weintraub and L. M. Reineke, J. Am. Chem. Soc., **75** 5768 (1953)

Cephalosporium subverticillatum (40%)

Bodánszky, A., J. Kollonitsch and G. Wix, Experientia, **11**, 384 (1955)

Cladosporium resinae

Fonken, G. S., H. C. Murray and L. M. Reineke,. J. Am. Chem. Soc., **82**, 5507 (1960)

Fusarium solani

Szpilfogel, S. A. et al., Rec. Trav. Chim., **75**, 402 (1956)

Gliocladium catenulatum ATCC 10523
Penicillium lilacinum

Peterson, D. H., S. H. Eppstein, P. D. Meister, H. C. Murray, H. M. Leigh, A. Weintraub and L. M. Reineke, J. Am. Chem. Soc., **75**, 5768 (1953)

Penicillium lilacinum

Sebek, O. K., L. M. Reineke and D. H. Peterson,. J. Bacteriol., **83**, 1327 (1962)

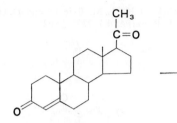

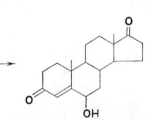

Progesterone

6β-Hydroxyandrost-4-ene-3, 17-dione

Gliocladium catenulatum

Peterson, D. H., S. H. Eppstein, P. D. Meister, H. C. Murray, H. M. Leigh, A. Weintraub and L. M. Reineke, J. Am. Chem. Soc., **75**, 5768 (1953)

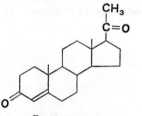

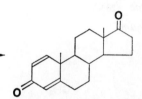

Progesterone

Androsta-1, 4-diene-3, 17-dione

Calonectria decora

Vischer, E., Ch. Meystre and A. Wettstein, Helv. Chim. Acta., **38**, 835 (1955)

Cylindrocarpon radicicola ATCC 11011

Peterson, G. E., R. W. Thoma, D. Perlman and J. Fried, J. Bacteriol., **74**, 684 (1957)

Fusarium caucasicum
Fusarium solani

Vischer, E. and A. Wettstein, Experientia, **9**, 371 (1953)

Fusarium solani

Nishikawa, M., S. Noguchi and T. Hasegawa, Pharm. Bull. (Japan), **3**, 322 (1955)

Streptomyces lavendulae strain Rutgers Univ. No. 3440-14 (7%)

Fried, J., R. W. Thoma and A. Klingsberg, J. Am. Chem. Soc., **75**, 5764 (1953)

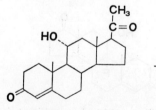

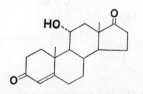

11α-Hydroxyprogesterone

11α-Hydroxyandrost-4-ene-3, 17-dione

Aspergillus chevalieri

Čapek, A., O. Hanč, K. Macek, M. Tadra and E. Riedl-Tůmová, Naturwiss, **43**, 471 (1956)

Aspergillus oryzae (20%)
Penicillium citrinum

Penicillium lilacinum

Sebek, O. K., L. M. Reineke and D. H. Peterson, J. Bacteriol., 83, 1327 (1962)

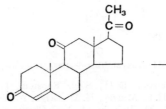 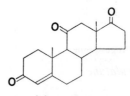

11-Oxoprogesterone   Adrenosterone

Aspergillus chevalieri

Aspergillus oryzae (20%)

Penicillium citrinum

Čapek, A., O. Hanč, K. Macek, M. Tadra and E. Riedl-Tůmová, Naturwiss., 43, 471 (1956)

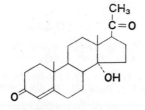 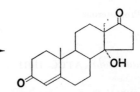

14α-Hydroxyprogesterone   14α-Hydroxyandrost-4-ene-3, 17-dione

Penicillium lilacinum

Peterson, D. H., S. H. Eppstein, P. D. Meister, H. C. Murray, H. M. Leigh, A. Weintraub and L. M. Reineke, J. Am. Chem. Soc., 75, 5768 (1953)

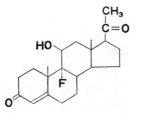 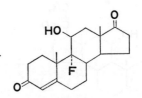

9α-Fluoro-11β-hydoxyprogesterone   9α-Fluoro-11β-hydroxyandrost-4-ene-3, 17-dione

Cylindrocarpon radicicola

U. S. Pat. 2,955,075

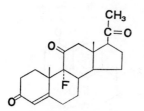

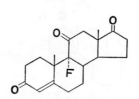

9α-Fluoro-11-oxoprogesterone

Cylindrocarpon radicicola

9α-Fluoroandrost-4-ene-3, 11, 17-trione

U. S. Pat. 2,955,075

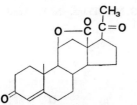

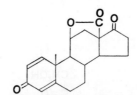

11β-Hydroxy-3, 20-dioxopregn-
4-en-18-oic acid-18, 11-lactone

Fusarium solani

11β-Hydroxy-3, 17-dioxoandrosta-1, 4-
dien-18-oic acid-18, 11-lactone

Urech, J., E. Vischer and A. Wettstein, Paper,
Meeting Swiss Chem. Soc., September (1961)

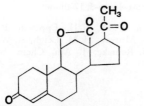

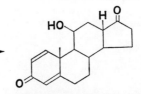

11β-Hydroxy-3, 20-dioxopregn-
4-en-18-oic acid-18, 11-lactone

Fusarium solani

11β-Hydroxy-18-norandrosta-1, 4-
diene-3, 17-dione

Urech, J., E. Vischer and A. Wettstein, Paper,
Meeting Swiss Chem. Soc., September (1961)

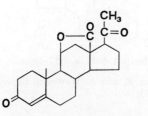

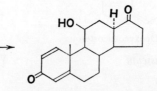

11β-Hydroxy-3, 20-dioxopregn
4-en-18-oic acid-18, 11-lactone

Fusarium solani

11β-Hydroxy-18-nor-18-isoandrosta-
1, 4-diene-3, 17-dione

Urech, J., E. Vischer and A. Wettstein, Paper,
Meeting Swiss Chem. Soc., September (1961)

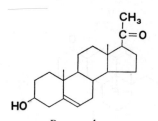

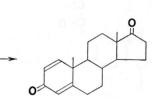

Pregnenolone

Androsta-1, 4-diene-3, 17-dione

Fusarium caucasicum
Fusarium solani

Vischer, E. and A. Wettstein, Experientia, **9**, 371 (1953)

Pycnodothis sp.

Shull, G. M., Trans. N. Y. Acad. Sci., **19**, 147 (1956)

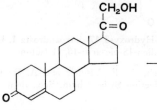

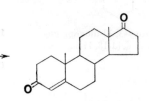

11-Deoxycorticosterone

Androst-4-ene-3, 17-dione

Aspergillus sp.
Penicillium sp.

Peterson, D. H., S. H. Eppstein, P. D. Meister, H. C. Murray, H. M. Leigh, A. Weintraub and L. M. Reineke, J. Am. Chem Soc., **75**, 5768 (1953)

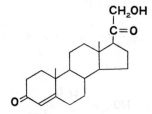

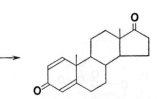

11-Deoxycorticosterone

Androsta-1, 4-diene-3, 17-dione

Fusarium caucasicum
Fusarium solani

Vischer, E. and A. Wettstein, Experientia, **9**, 371 (1953)

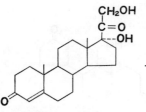

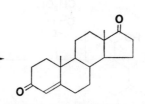

11-Deoxycortisol

Androst-4-ene-3, 17-dione

Didymella lycopersici, conidia

Penicillium chrysogenum, conidia

Vezina, C., S.N. Sehgal and K. Singh, Appl. Microbiol., **11**, 50 (1963)

Pseudomonas chlororaphis IAM 1511

Naito, A., Y. Sato, H. Iizuka and K. Tsuda, Steroids, **3**, 327 (1964)

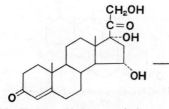

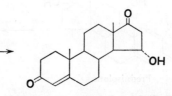

15α-Hydroxy-11-deoxycortisol

15α-Hydroxyandrost-4-ene-3, 17-dione

Pseudomonas chlororaphis IAM 1511

Naito, A., Y. Sato, H. Iizuka and K. Tsuda, Steroids, **3**, 327 (1964)

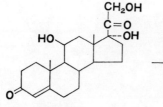

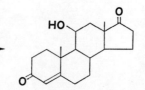

Cortisol

11β-Hydroxyandrost-4-ene-3, 17-dione

Pseudomonas chlororaphis IAM 1511

Naito, A., Y. Sato, H. Iizuka and K. Tsuda, Steroids, **3**, 327 (1964)

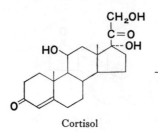
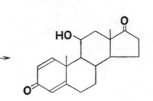

Cortisol   11β-Hydroxyandrosta-1, 4-diene-
3, 17-dione

Pseudomonas chlororaphis IAM 1511   Naito, A., Y. Sato, H. Iizuka and K. Tsuda,
Steroids, **3**, 327 (1964)

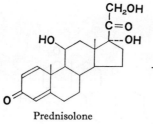
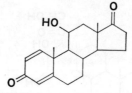

Prednisolone   11β-Hydroxyandrosta-1, 4-diene-
3, 17-dione

Pseudomonas chlororaphis IAM 1511   Naito, A., Y. Sato, H. Iizuka and K. Tsuda,
Steroids, **3**, 327 (1964)

E. Lactone Formation

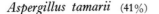

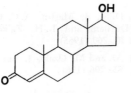 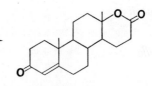

Testosterone   Testololactone

Aspergillus tamarii (41%)

Brannon, D. R., J. Martin, A. C. Oehlschlager, N. N. Durham and L. H. Zalkow, J. Org. Chem., **30**, 760 (1965)

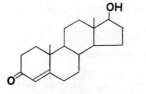 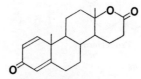

Testosterone   1-Dehydrotestololactone

Cylindrocarpon radicicola ATCC 11011
(50%)

Fried, J., R. W. Thoma and A. Klingsberg, J. Am. Chem. Soc., **75**, 5764 (1953)

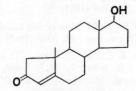 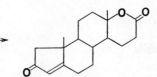

A-Nortestosterone   A-Nortestololactone

Penicillium citrinum

U. S. Pat. 2,998,428

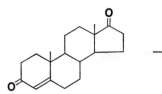
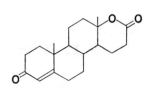

Androst-4-ene-3, 17-dione

Testololactone

Aspergillus tamarii

Brannon, D. R., J. Martin, A. C. Oehlschlager, N. N. Durham and L. H. Zalkow, J. Org. Chem., **30**, 760 (1965)

Pythium ultimum

Shirasaka, M. and M. Ozaki, J. Agr. Chem. Soc. (Japan), **35**, 206 (1961)

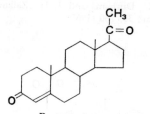
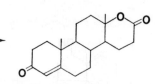

Progesterone

Testololactone

Aspergillus flavus

Fried, J., R. W. Thoma and A. Klingsberg, J. Am. Chem. Soc., **75**, 5764 (1953)

Aspergillus oryzae

Čapek, A., O. Hanč, K. Macek, M. Tadra and E. Riedl-Tůmová, Naturwiss., **43**, 471 (1956)

Aspergillus tamarii (70%)

Brannon, D. R., J. Martin, A. C. Oehlschlager, N. N. Durham and L. H. Zalkow, J. Org. Chem., **30**, 760 (1965)

Cephalosporium subverticillatum (54%)

Bodánszky, A., J. Kollonitsch and G. Wix, Experientia, **11**, 384 (1955)

Cladosporium resinae

Fonken, G. S., H. C. Murray and L. M. Reineke, J. Am. Chem. Soc., **82**, 5507 (1960)

Collybia dryophila C-59

Schuytema, E. C., M. P. Hargie, D. J. Siehr, I. Merits, J. R. Schenck, M. S. Smith and E. L. Varner, Appl. Microbiol., **11**, 256 (1963)

Penicillium adametzi

Peterson, D. H., S. H. Eppstein, P. D. Meister, H. C. Murray, H. M. Leigh, A. Weintraub and L. M. Reineke, J. Am. Chem. Soc., **75**, 5768 (1953)

Penicillium chrysogenum (70%)

Fried, J., R. W. Thoma and A. Klingsberg, J. Am. Chem. Soc., **75**, 5764 (1953)

Penicillium lilacinum

Sebek, O. K., L. M. Reineke and D. H. Peterson, J. Bacteriol., **83**, 1327 (1962)

Pythium ultimum

Shirasaka, M. and M. Ozaki, J. Agr. Chem. Soc. (Japan), **35**, 206 (1961)

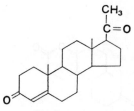

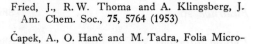

Progesterone

1-Dehydrotestololactone

Cylindrocarpon radicicola ATCC 11011
(50%)

Fusarium lateritium (40%)

Fusarium solani

Fried, J., R. W. Thoma and A. Klingsberg, J. Am. Chem. Soc., **75**, 5764 (1953)

Čapek, A., O. Hanč and M. Tadra, Folia Microbiol., **8**, 120 (1963)

Nishikawa, M., S. Noguchi and T. Hasegawa, Pharm. Bull. (Japan), **3**, 322 (1955)

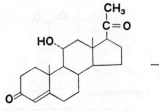

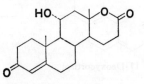

11α-Hydroxyprogesterone

11α-Hydroxytestololactone

Penicillium lilacinum

Sebek, O. K., L. M. Reineke and D. H. Peterson, J. Bacteriol., **83**, 1327 (1962)

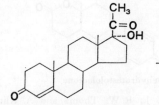

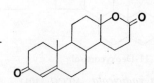

17α-Hydroxyprogesterone

Testololactone

Aspergillus flavus

Peterson, D. H., S. H. Eppstein, P. D. Meister, H. C. Murray, H. M. Leigh, A. Weintraub and L. M. Reineke, J. Am. Chem. Soc., **75**, 5768 (1953)

Pythium ultimum

Shirasaka, M. and M. Ozaki, J. Agr. Chem. Soc. (Japan), **35**, 206 (1961)

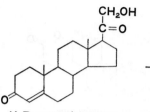

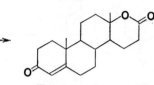

11-Deoxycorticosterone	Testololactone

Fusarium sp.

Shull, G. M., Trans. N. Y. Acad. Sci., **19**, 147 (1956)

Pythim ultimum

Shirasaka, M. and M. Ozaki, J. Agr. Chem. Soc. (Japan), **35**, 206 (1961)

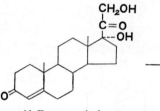

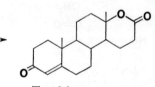

11-Deoxycortisol	Testololactone

Curvularia lunata

Shull, G. M., Trans. N. Y. Acad. Sci., **19**, 147 (1956)

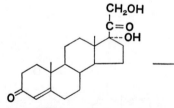

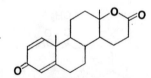

11-Deoxycortisol	1-Dehydrotestololactone

Cylindrocarpon radicicola ATCC 11011 (50%)

Fried, J., R. W. Thoma and A. Klingsberg, J. Am. Chem. Soc., **75**, 5764 (1953)

Pseudomonas chlororaphis IAM 1511

Naito, A., Y. Sato, H. Iizuka and K. Tsuda, Steroids, **3**, 327 (1964)

F.   AROMATIZATION

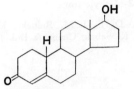

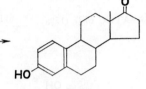

19-Nortestosterone

Estrone

Bacillus sphaericus   ATCC 7055

Nocardia corallina

Gaul, C., R. I. Dorfman and S. R. Stitch, Bio. chem. Biophys. Acta, **49** 387 (1961)

U. S. Pat. 3,087,864

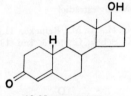

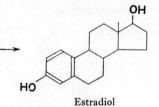

19-Nortestosterone

Estradiol

Arthrobacter simplex

Japan Pat. 443,426

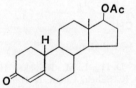

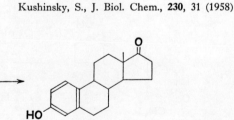

19-Nortestosterone 17-acetate

Estradiol 17-acetate

Corynebacterium simplex   ATCC 6946

Kushinsky, S., J. Biol. Chem., **230**, 31 (1958)

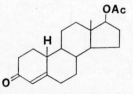

19-Nortestosterone 17-acetate

Estrone

Pseudomonas testosteroni   ATCC 11996

Levy, H. R. and P. Talalay, J. Am. Chem. Soc., **79**, 2658 (1957)

— 183 —

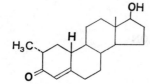
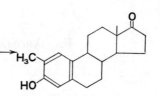

2α-Methyl-19-nortestosterone

2-Methylestrone

Septomyxa affinis ATCC 6737 (17%)

Peterson, D. H., L. M. Reineke, H. C. Murray and O. K. Sebek, Chem. & Ind. 1301 (1960)

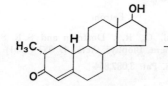
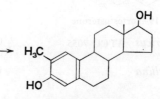

2α-Methyl-19-nortestosterone

2-Methylestradiol

Septomyxa affinis ATCC 6737

Peterson, D. H., L. M. Reineke, H. C. Murray and O. K. Sebek, Chem. & Ind. 1301 (1960)

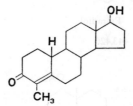
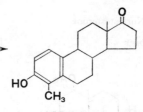

4-Methyl-19-nortestosterone

4-Methylestrone

Septomyxa affinis ATCC 6737 (12%)

Peterson, D. H., L. M. Reineke, H. C. Murray and O. K. Sebek, Chem. & Ind. 1301 (1960)

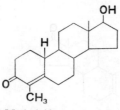
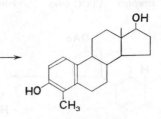

4-Methyl-19-nortestosterone

4-Methylestradiol

Septomyxa affinis ATCC 6737

Peterson, D. H., L. M. Reineke, H. C. Murray and O. K. Sebek, Chem. & Ind. 1301 (1960)

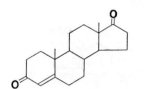

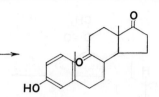

Androst-4-ene-3,17-dione

9,10-Seco-3-hydroxyandrosta-
1,3,5(10)-triene-9,17-dione

Arthrobacter sp. B-22-8 ATCC 13260

Dodson, R. M. and R. D. Muir, J. Am. Chem. Soc., **83**, 4627 (1961)

Nocardia restrictus

Sih, C. J., Biochem. Biophys. Res. Comm. **7**, 87 (1962)

Pseudomonas sp. B-20-184

Dodson, R. M. and R. D. Muir, J. Am. Chem. Soc., **83**, 4627 (1961)

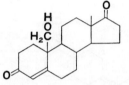

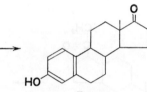

19-Hydroxyandrost-4-ene-3,17-dione

Estrone

Pseudomonas sp. B-20-184

Dodson, R. M. and R. D. Muir, J. Am. Chem. Soc., **83**, 4627 (1961)

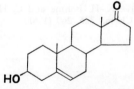

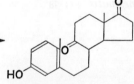

Dehydroepiandrosterone

9,10-Seco-3-hydroxyandrosta-
1,3,5(10)-triene-9,17-dione

Mycobacterium smegmatis SG 98

Schubert, K., K.-H. Böhme and C. Hörhold, Z. Naturforsch., **15 B**, 584 (1960)

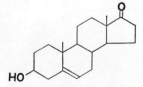

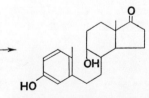

Dehydroepiandrosterone

9,10-Seco-3,9-dihydroxyandrosta-
1,3,5(10)-trien-17-one

Mycobacterium smegmatis SG 98

Schubert, K., K.-H. Böhme and C. Hörhold, Z. Naturforsch., **15 B**, 584 (1960)

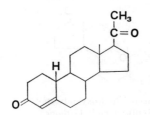

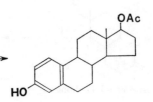

19-Norprogesterone

17β-Acetoxyestra-1, 3, 5(10)-
trien-3-ol

Corynebacterium simplex ATCC 6946

Bowers, A., C. Casas-campillo and C. Djerassi,
Tetrahedron, **2**, 165 (1958)

Streptomyces lavendulae ATCC 8664

Gaul, C., R. I. Dorfman and S. R. Stitch, Biochem. Biophys. Acta., **49**, 387 (1961)

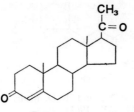

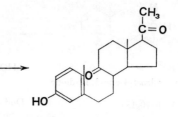

Progesterone

9, 10-Seco-3-hydroxypregna-
1, 3, 5(10)-triene-9, 20-dione

Mycobacterium smegmatis SG 98

Schubert, K., K.-H. Böhme and C. Hörhold, Z.
Physiol. Chem., **325**, 260 (1961)

G. ISOMERIZATION

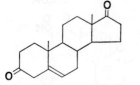

Androst-5-ene-3, 17-dione

Pseudomonas sp.

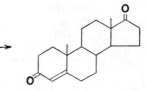

Androst-4-ene-3, 17-dione

Talalay, P. and V. S. Wang, Biochem. Biophys. Acta, **18**, 300 (1955)

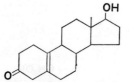

17β-Hydroxy-19-norandrost-
5(10)-en-3-one

Pseudomonas sp.

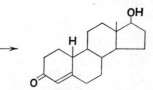

17β-Hydroxy-19-norandrost-4-
en-3-one (19-Nortestosterone)

Talalay, P. and V. S. Wang, Biochem. Biophys. Acta, **18**, 300 (1955)

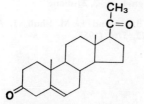

Pregn-5-ene-3, 20-dione

Pseudomonas sp.

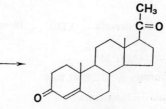

Progesterone

Talalay, P. and V. S. Wang, Biochem. Biophys, Acta, **18**, 300 (1955)

H. Epoxidation

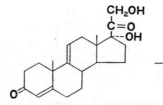

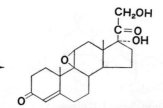

9(11)-Dehydro-11-deoxycortisol

17α, 21-Dihydroxy-9β, 11β-oxidopregn-
4-ene-3, 20-dione

Cunninghamella blakesleeana ATCC
9245

Curvularia lunata NRRL 2380

Bloom, B. M. and G. M. Shull, J. Am. Chem.
Soc., **77**, 5767 (1955)

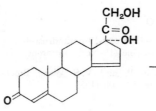

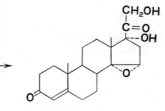

14(15)-Dehydro-11-deoxycortisol

17α, 21-Dihydroxy-14α, 15α-oxido-
pregn-4-ene-3, 20-dione

Cunninghamella blakesleeana ATCC
9245

Curvularia lunata NRRL 2380

Helicostylum piriforme ATCC 8992, 8686

Mucor griseocyanus ATCC 1207a

Mucor parasiticus ATCC 6476

Bloom, B. M. and G. M. Shull, J. Am. Chem.
Soc., **77**, 5767 (1955)

I. Hydrolysis

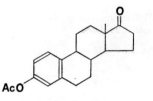

Estrone 3-acetate

Baker's yeast

'α'-Estradiol

Mamoli, L., Ber., **71 B**, 2696 (1938)

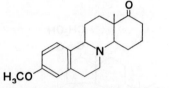

8-Aza-D-homoestrone 3-methylether

Aspergillus flavus

8-Aza-D-homoestrone

Curtis, P. J., Biochem. J., **97**, 148 (1965)

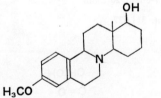

8-Aza-D-homoestradiol 3-methylether

Cunninghamella blakesleeana

8-Aza-D-homoestradiol

Curtis, P. J., Biochem. J., **97**, 148 (1965)

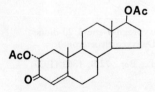
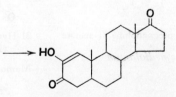

2α-Hydroxytestosterone
2α, 17β-diacetate

Nocardia carollina

2-Hydroxyandrost-1-ene-3, 17-dione

U. S. Pat. 3,087,864

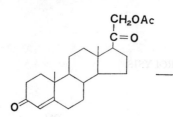

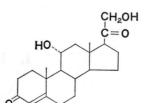

11-Deoxycorticosterone 21-acetate

11α, 21-Dihydroxypregn-4-ene-3, 20-dione

Aspergillus clavatus
Aspergillus fischeri
Aspergillus nidulans (15.5%)
Aspergillus ustus

U. S. Pat. 2,649,402

Rhizopus nigricans

U. S. Pat. 2,602,769

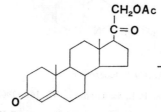

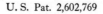

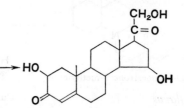

11-Deoxycorticosterone 21-acetate

2β, 15β, 21-Trihydroxypregn-4-ene-3, 20-dione

Sclerotinia sclerotiorum

Japan Pat. 311,627

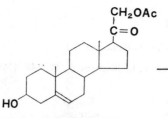

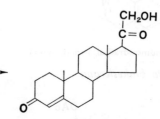

21-Hydroxypregnenolone 21-acetate

21-Hydroxypregn-4-ene-3, 20-dione
(11-Deoxycorticosterone)

Corynebacterium mediolanum

Mamoli, L., Ber., **72 B**, 1863 (1939)

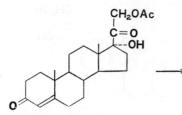

11-Deoxycortisol 21-acetate

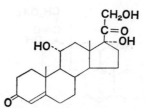

11α, 17α, 21-Trihydroxypregn-
4-ene-3, 20-dione

Dactylium dendroides

Dan. Pat. 94,041

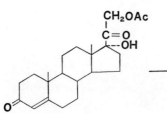

11-Deoxycortisol 21-acetate

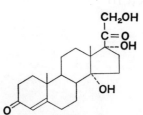

14α, 17α, 21-Trihydroxypregn-
4-ene-3, 20-dione

Mycobacterium lacticola
Mycobacterium smegmatis

Belg. Pat. 538,327

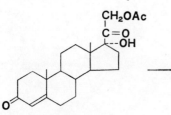

11-Deoxycortisol 21-acetate

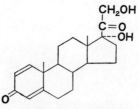

17α, 21-Dihydroxypregna-1, 4-
diene-3, 20-dione

Mycobacterium lacticola
Mycobacterium smegmatis

Belg. Pat. 538,327

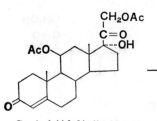 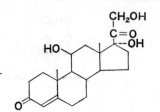

Cortisol 11β, 21-diacetate

11β, 17α, 21-Trihydroxypregn-
4-ene-3, 20-dione
(Cortisol)

Flavobacterium dehydrogenans var. *hydrolyticum* SCH 111

Charney, W., L. Weber and E. Oliveto, Arch.
Biochem. Biophys., **79**, 402 (1959)

J. ESTERIFICATION

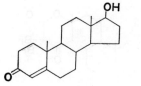

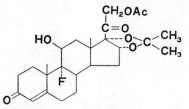

Testosterone   Testosterone 17-acetate

Saccharomyces fragilis ATCC 10022   McGuire, J, S., E. S. Maxwell and G. M. Tom-
                                      kins, Biochem. Biophys. Acta, **45**, 392 (1960)

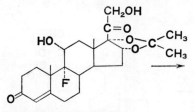

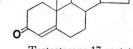

9α-Fluoro-11β, 21-dihydroxy   9α-Fluoro-11β, 21-dihydroxy-16α, 17α-
16α, 17α-isopropylidenedioxy-   isopropylidenedioxypregn-4-
pregn-4-ene-3, 20-dione   ene-3, 20-dione 21-acetate

Trichoderma glaucum Lederle culture Holmlund, C. E., L. l. Feldman, N. E. Rigler,
No. Z-696   B. E. Nielsen and R. H. Evans Jr., J. Am.
                                      Chem. Soc., **83**, 2586 (1961)

K. HALOGENATION

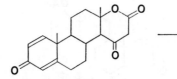

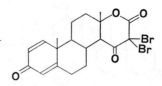

15-Oxo-1-dehydrotestololactone

16-Dibromo-15-Oxo-1-dehydro-
testololactone

Caldariomyces fumago ATCC 16373

Neidleman, S. L., P. A. Diassi, B. Junta, R. M.
Palmere and S. C. Pan, Tetrahedron Letter
No. 44, 5337 (1966)

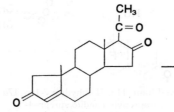

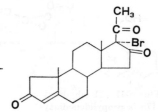

16-Oxo-A-norprogesterone

17α-Bromo-A-norpregn-4-
ene-3, 16, 20-trione

Caldariomyces fumago ATCC 16373

Neidleman, S. L., P. A. Diassi, B. Junta, R. M.
Palmere and S. C. Pan, Tetrahedron Letter
No. 44, 5337 (1966)

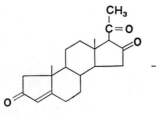

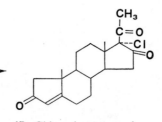

16-Oxo-A-norprogesterone

17α-Chloro-A-norpregn-4-
ene-3, 16, 20-trione

Caldariomyces fumago ATCC 16373

Neidleman, S. L., P. A. Diassi, B. Junta, R. M.
Palmere and S. C. Pan, Tetrahedron Letter
No. 44, 5337 (1966)

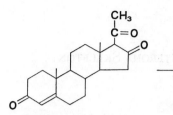

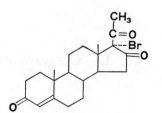

16-Oxoprogesterone

Caldariomyces fumago ATCC 16373 (50%)

17α-Bromopregn-4-ene-3, 16, 20-trione

Neidleman, S. L., P. A. Diassi, B. Junta, R. M. Palmere and S. C. Pan, Tetrahedron Letter No. 44, 5337 (1966)

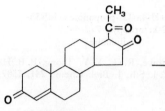

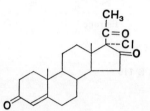

16-Oxoprogesterone

Caldariomyces fumago ATCC 16373 (50%)

17α-Chloropregn-4-ene-3, 16, 20-trione

Neidleman, S. L., P. A. Diassi, B. Junta, R. M. Palmere and S. C. Pan, Tetrahedron Letter No. 44, 5337 (1966)

L. Cleavage of Steroid Skeleton

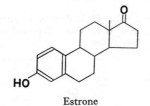

Estrone

Nocardia sp. E 110

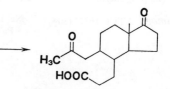

3aα-H-4α-[3′-propanoic acid]-5β-
[2-ketopropyl]-7aβ-methyl-
1-indanone

Coombe, R. G., Y. Y. Tsong, P. B. Hamilton and
C. J. Sih, J. Biol. Chem., **241**, 1587 (1966)

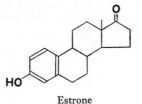

Estrone

Nocardia sp. E 110

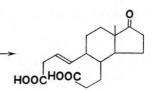

3aα-H-4α-[3′-propanoic acid]-5β-
[4′-but-3-enoic acid]-7aβ-
methyl-1-indanone

Coombe, R. G., Y. Y. Tsong, P. B. Hamilton and
C. J. Sih, J. Biol. Chem., **241**, 1587 (1966)

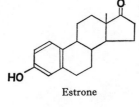

Estrone

Nocardia sp. E 110

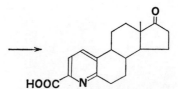

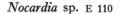

2-Carboxy-7aβ-methyl-7-keto-9aα-H-
indano-[5, 4f]-5aα, 10, 10aβ, 11-
tetrahydroquinoline

Coombe, R. G., Y. Y. Tsong, P. B. Hamilton and
C. J. Sih, J. Biol. Chem., **241**, 1587 (1966)

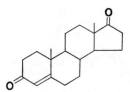

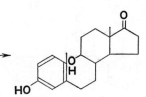

Androst-4-ene-3, 17-dione

9, 10-Seco-3, 9-dihydroxyandrosta-
1, 3, 5(10)-trien-17-one

Mycobacterium smegmatis SG 98

Schubert, K., K.-H. Böhme and C. Hörhold, Z.
Naturforsch., **15B**, 584 (1960)

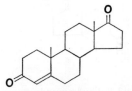

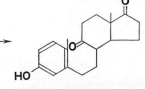

Androst-4-ene-3, 17-dione

9, 10-Seco-3-hydroxyandrosta-
1, 3, 5(10)-triene-9, 17-dione

Arthrobacter sp. B-22-8 ATCC 13260

Dodson, R. M. and R. D. Muir, J. Am. Chem.
Soc., **83**, 4627 (1961)

Mycobacterium smegmatis SG 98

Schubert, K., K.-H. Böhme and C. Hörhold, Z.
Naturforsch., **15B**, 584 (1960)

Nocardia restrictus

Sih, C. J., Biochem. Biophys. Res. Comm., **7**, 87
(1962)

Pseudomonas sp. B-20-184

Dodson, R. M. and R. D. Muir, J. Am. Chem.
Soc., **83**, 4627 (1961)

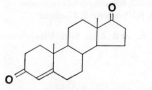

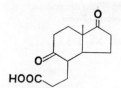

Androst-4-ene-3, 17-dione

7aβ-Methylperhydroindane-1, 5-dione-
4α-[3'-propionic acid]

Nocardia restrictus ATCC 14887

Sih, C. J., S. S. Lee, Y. Y. Tsong and K. C.
Wang, J. Am. Chem. Soc., **87**, 1385 (1965)

Nocardia restrictus No. 545

Sih, C. J. and K. C. Wang, J. Am. Chem. Soc.,
85, 2135 (1963)

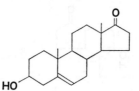 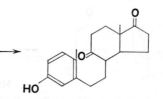

Dehydroepiandrosterone

9, 10-Seco-3-hydroxyandrosta-
1, 3, 5(10)-triene-9, 17-dione

Mycobacterium smegmatis SG 98

Schubert, K., K.-H. Böhme and C. Hörhold, Z
Naturforsch., **15B** 584 (1960)

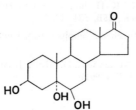 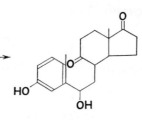

Dehydroepiandrosterone

9, 10-Seco-3, 9-dihydroxyandrosta-
1, 3, 5(10)-trien-17-one

Mycobacterium smegmatis SG 98

Schubert, K., K.-H. Böhme and C. Hörhold, Z
Naturforsch., **15B** 584 (1960)

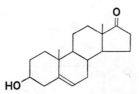 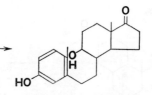

3β, 5α, 6β-Trihydroxy-
androstan-17-one

9, 10-Seco-3, 6R-dihydroxyandrosta-
1, 3, 5(10)-triene-9, 17-dione

Nocardia restrictus No. 545

Lee, S. S. and C. J. Sih, Biochemistry, **3**, 1267
(1964)

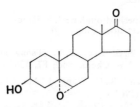 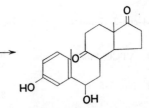

3β-Hydroxy-5α, 6α-oxido-
androstan-17-one

9, 10-Seco-3, 6S-dihydroxyandrosta-
1, 3, 5(10)-triene-9, 17-dione

Nocardia restrictus No. 545

Lee, S. S. and C. J. Sih, Biochemistry, **3**, 1267
(1964)

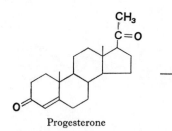

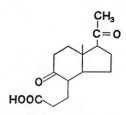

Progesterone

7a-Methyl-1-acetylperhydroindanone-
(5)-[β-propionic acid-(4)]

Mycobacterium smegmatis SG 98

Schubert, K., K.-H. Böhme and C. Hörhold,
Hoppe-Seyler's, Z. Physiol. Chem., **325**, 260
(1961)

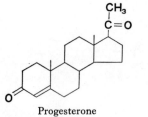

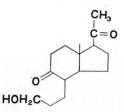

Progesterone

7a-Methyl-1-acetylperhydroindanone-
(5)-[β-propylalcohol-(4)]

Mycobacterium smegmatis SG 98

Schubert, K., K.-H. Böhme and C. Hörhold,
Steroids, **4**, 581 (1964)

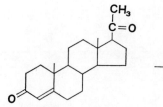

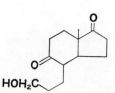

Progesterone

7a-Methylperhydroindanedione-
(1, 5)-[β-propylalcohol-(4)]

Mycobacterium smegmatis SG 98

Schubert, K., K.-H. Böhme and C. Hörhold,
Steroids, **4**, 581 (1964)

II. MICROBIAL TRANSFORMATION
OF BILE ACIDS

Structures of typical bile acids

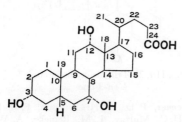

Cholic acid
(3α, 7α, 12α-Trihydroxy-5β-cholanic acid)

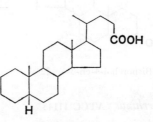

5β-Cholanic acid

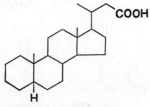

24-Nor-5β-cholanic acid

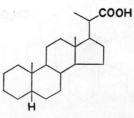

23, 24-Bisnor-5β-cholanic acid

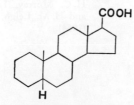

Etiocholanic acid
(5β-Androstane-17β-carboxylic acid)

A. HYDROXYLATION

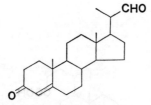

Bisnorchol-4-en-3-on-22-al

Rhizopus arrhizus ATCC 11145
Rhizopus nigricans ATCC 6227b

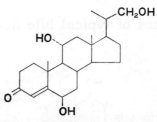

6β, 11α, 22-Trihydroxybisnorchol-
4-en-3-one

Meister, P. D., D. H. Peterson, S. H. Eppstein,
H. C. Murray, L. M. Reineke, A. Weintraub
and H. M. Leigh, J. Am. Chem. Soc., **76**, 5679
(1954)

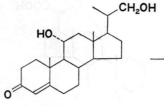

11α, 22-Dihydroxybisnorchol-4-
en-3-one

Cunninghamella blakesleeana ATCC
8688a

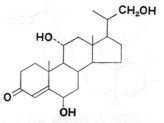

6β, 11α, 22-Trihydroxybisnorchol-
4-en-3-one

Meister, P. D., D. H. Peterson, S. H. Eppstein,
H. C. Murray, L. M. Reineke, A. Weintraub
and H. M. Leigh, J. Am. Chem. Soc., **56**, 5679
(1954)

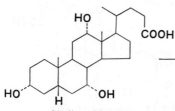

Cholic acid

Alcaligenes faecalis

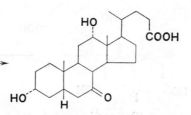

3α, 12α-Dihydroxy-7-oxocholanic acid

Hoehn, W. M., L. H. Schmidt and H. B. Hughes,
J. Biol. Chem., **152**, 59 (1944)

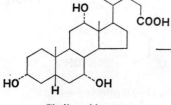

Cholic acid

Streptomyces gelaticus strain 1164

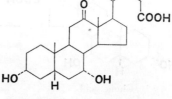

3α, 7α-Dihydroxy-12-oxocholanic acid

Hayakawa, S., Y. Saburi and H. Teraoka, Proc.
Japan. Acad., **32**, 519 (1956)

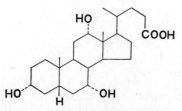

Cholic acid

Streptomyces gelaticus strain 1164 (6.7%)

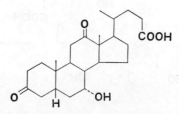

7α-Hydroxy-3, 12-dioxocholanic acid

Hayakawa, S., Y. Saburi and H. Teraoka, Proc.
Japan Acad., **32**, 519 (1956)

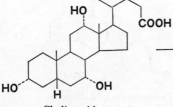

Cholic acid

Alcaligenes faecalis

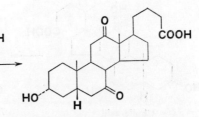

3α-Hydroxy-7, 12-dioxocholanic acid

Hoehn, W. M., L. H. Schmidt and H. B. Hughes,
J. Biol. Chem., **152**, 59 (1944)

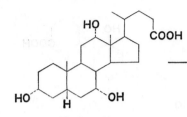

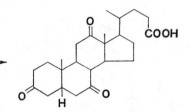

Cholic acid

3, 7, 12-Trioxocholanic acid

Alcaligenes faecalis (83%)

Schmidt, L. H., H. B. Hughes, M. H. Green and E. Cooper, J. Biol. Chem., **145**, 229 (1942)

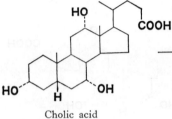

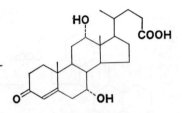

Cholic acid

7α, 12α-Dihydroxy-3-oxo-chol-4-enic acid

A soil bacterium   strain CE-1

Eguchi, T., J. Biochem. (Japan), **44**, 81 (1957)

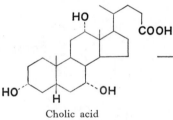

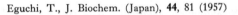

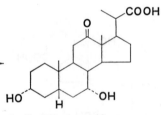

Cholic acid

3α, 7α-Dihydroxy-12-oxo-bisnorcholanic acid

Streptomyces gelaticus   strain 1164

Hayakawa, S., Y. Saburi and I. Akaeda, J. Biochem. (Japan), **44**, 109 (1957)

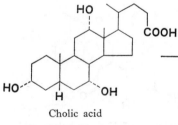

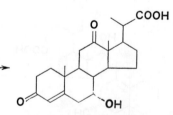

Cholic acid

7α-Hydroxy-3, 12-dioxobis-norchol-4-enic acid

Streptomyces gelaticus   strain 1164

Hayakawa, S., Y. Saburi, T. Fujii and Y. Sonoda, J. Biochem. (Japan), **43**, 723 (1956)

B. DEHYDROGENATION

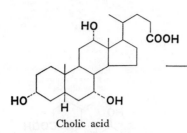

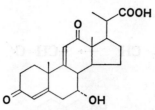

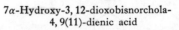

Cholic acid

7α-Hydroxy-3, 12-dioxobisnorchola-
4, 9(11)-dienic acid

Actinomyces No. 1164 (6.7%)

Hayakawa, S., Proc. Japan Acad., **30**, 133 (1954)

(b) 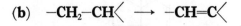 $-CH_2-CH\langle \longrightarrow -CH=C\langle$

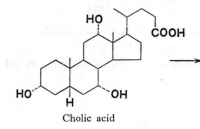

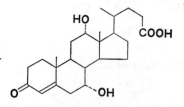

Cholic acid   7α, 12α-Dihydroxy-3-oxochol-
4-enic acid

A soil bacterium CE-1   Eguchi, T., J. Biochem. (Japan), **44**, 81 (1957)

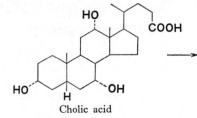

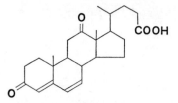

Cholic acid   3, 12-Dioxochola-4, 6-dienic acid

Streptomyces gelaticus strain 1164   Hayakawa, S., Y. Saburi and H. Teraoka, Proc.
Japan Acad., **32**, 519 (1956)

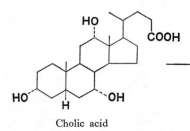

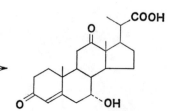

Cholic acid   7α-Hydroxy-3, 12-dioxobisnorchol-
4-enic acid

Streptomyces gelaticus strain 1164   Hayakawa, S., Y. Saburi, T. Fujii and Y. Sonoda, J. Biochem. (Japan), **43**, 723 (1956)

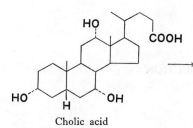

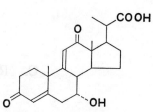

Cholic acid

Actinomyces sp. (7.3%)

7α-Hydroxy-3, 12-dioxobisnorchola-
4, 9(11)-dienic acid

Hayakawa, S., Proc. Japan Acad., **30**, 133 (1954)

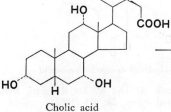

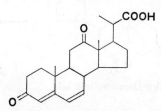

Cholic acid

3, 12-Dioxobisnorchola-4, 6-dienic acid

Streptomyces gelaticus strain 1164

Hayakawa, S., Y. Saburi and I. Akaeda, J. Biochem. (Japan), **44**, 109 (1957)

C. REDUCTION

(a) $\rangle$C=O $\longrightarrow$ $\rangle$CH–OH

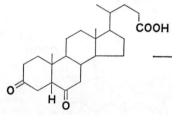

3, 6-Dioxocholanic acid

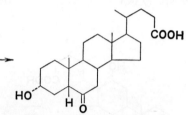

3α-Hydroxy-6-oxocholanic acid

Yeast

Ercoli, A. and P. De Ruggieri, Boll. Soc. Ital. Biol. Sper., **28**, 611 (1952)

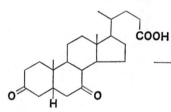

3, 7-Dioxocholanic acid

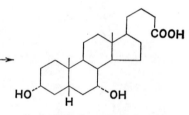

3α, 7α-Dihydroxycholanic acid

Escherichia coli

Sihn, T. S., J. Biochem. (Japan), **28**, 165 (1938)

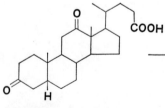

3, 12-Dioxocholanic acid

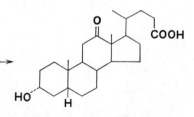

3α-Hydroxy-12-oxocholanic acid

Press yeast (11%)

Kim, C. H., Enzymologia, **4**, 119 (1937)

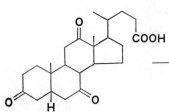

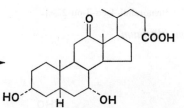

3, 7, 12-Trioxocholanic acid

7α-Hydroxy-3, 12-dioxocholanic acid

Escherichia coli

Fukui, T., J. Biochem. (Japan), **25**, 61 (1937)

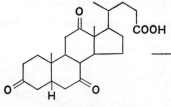

3, 7, 12-Trioxocholanic acid

3α, 7α-Dihydroxy-12-oxocholanic acid

Bacillus coli

Machida, M., J. Biochem. (Japan), **40**, 435 (1953)

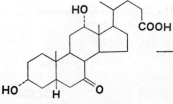

3α, 12α-Dihydroxy-7-oxo-
cholanic acid

Cholic acid

Escherichia coli

Machida, M., J. Med. Sci., **2**, 291 (1953)

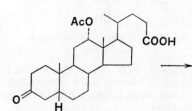

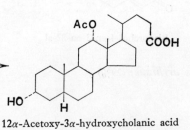

12α-Acetoxy-3-oxocholanic acid

12α-Acetoxy-3α-hydroxycholanic acid

Beer yeast (33%)

Kim, C. H., Enzymologia, **6**, 105 (1939)

(b) −CHO ⟶ −CH₂OH

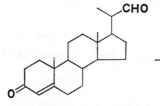 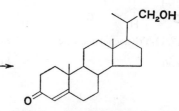

Bisnorchol-4-en-3-on-22-al   22-Hydroxybisnorchol-4-en-3-one

Penicillium lilacinum

Peterson, D. H., Record Chem. Progr., **17**, 211 (1956)

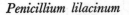

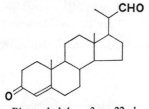 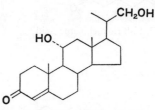

Bisnorchol-4-en-3-on-22-al   11α, 22-Dihydroxybisnorchol-4-en-3-one

Rhizopus nigricans (17.4%)

U. S. Pat. 2,602,769

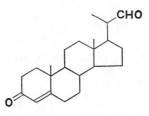 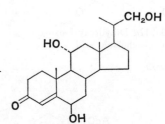

Bisnorchol-4-en-3-on-22-al   6β, 11α, 22-Trihydroxybisnorchol-4-en-3-one

Rhizopus arrhizus (29%)

Meister, P. D., D. H. Peterson, S. H. Eppstein, H. C. Murray, L. M. Reineke, A. Weintraub and H. M. Leigh, J. Am. Chem. Soc., **76**, 5679 (1954)

(c) $-CHOH- \longrightarrow -CH_2-$

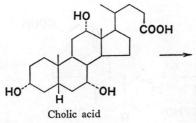

Cholic acid

Intestinal bacteria

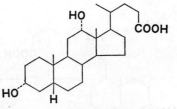

$3\alpha, 12\alpha$-Dihydroxycholanic acid

Baumgärtel et al., Deut. Z. Verdauungs-u. Stoff-
wechselkrankh., **11**, 257 (1951)

D. SIDE CHAIN DEGRADATION

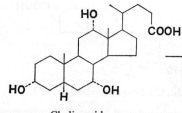 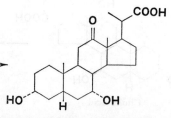

Cholic acid

3α, 7α-Dihydroxy-12-oxobisnor-
cholanic acid

Streptomyces gelaticus strain 1164

Hayakawa, S., Y. Saburi and I. Akaeda, J. Biochem. (Japan), **44**, 109 (1957)

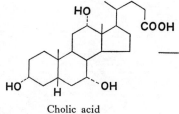 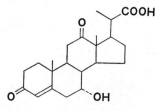

Cholic acid

7α-Hydroxy-3, 12-dioxobisnorchol-
4-enic acid

Streptomyces gelaticus strain 1164

Hayakawa, S., Y. Saburi, T. Fujii and Y. Sonoda.
J. Biochem. (Japan), **43**, 723 (1956)

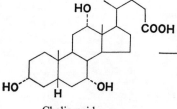 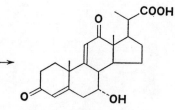

Cholic acid

7α-Hydroxy-3, 12-dioxobisnorchola-
4, 9(11)-dienic acid

Actinomyces sp. No. 1164 (6.7%)

Hayakawa, S., Proc. Japan Acad., **30**, 133 (1954)

III. MICROBIAL TRANSFORMATION OF STEROLS

Structures of typical sterols

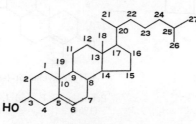

Cholesterol
(Cholest-5-en-3β-ol)

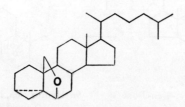

β-Sitosterol
(β-Sitost-5-en-3β-ol)

3α, 5-Cyclo-6β, 19-oxido-5α-cholestane

A. HYDROXYLATION

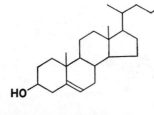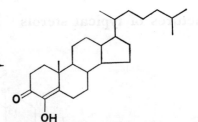

Cholesterol | 4-Hydroxycholest-4-en-3-one

Streptomyces sp.

Peterson, G. E. and J. R. Davis, Steroids, **4**, 677 (1964)

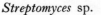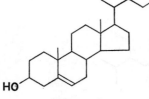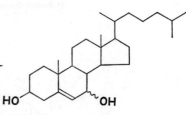

Cholesterol | 7ξ-Hydroxycholesterol

Proactinomyces roseus

Krámli, A. and J. Horváth, Nature, **162**, 619 (1948)

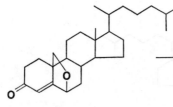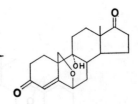

6β, 19-Oxidocholest-4-en-3-one | 9α-Hydroxy-6β, 19-oxido-androst-4-ene-3, 17-dione

CSD-10

Sih, C. J., S. S. Lee, Y. Y. Tsong, K. C. Wang and F. N. Chang, J. Am. Chem. Soc., **87**, 2765 (1965)

B. Dehydrogenation

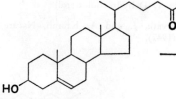 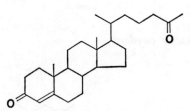

3-Hydroxynorcholest-5-en-25-one   Norcholest-4-en-3, 25-dione

Flavobacterium dehydrogenans (14%)

(*Micococcus dehydrogenans*)

Ercoli, A., Boll. Sci. Facolata Chim. Ind. Bologna 279 (1940)

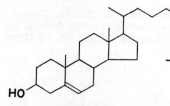 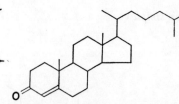

Cholesterol   Cholest-4-en-3-one

Arthrobacter simplex (30~40%)

Nagasawa, M., T. Hai, G. Tamura and K. Arima, Abstr. paper (39th and 40th Meeting Agr. Chem. Soc., Japan) p. 81, No. 415, 416 (1964) p. 63, No. 366 (1965)

Azotobacter sp.

Horváth, J. and A. Krámli, Nature, **160**, 639 (1947)

Flavobacterium maris (11~13%)

Arnaudi, C. and C. Colla, Experientia, **5**, 120 (1949)

Mycobacterium cholesterolicum

Stadtman, T. C., A. Cherkes and C. B. Anfinsen, J. Biol. Chem., **206**, 511 (1954)

Nocardia restrictus ATCC 14887

Sih, C. J. and K. C. Wang, J. Am. Chem. Soc., **87**, 1387 (1965)

Proactinomyces erythropolis

Turfitt, G. E., Biochem. J., **42**, 376 (1948)

Proactinomyces roseus

Krámli, A. and J. Horváth, Nature, **162**, 619 (1948)

Streptomyces sp.

Peterson, G. E. and J. R. Davis, Steroids, **4**, 677 (1964)

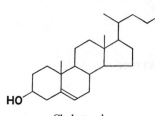   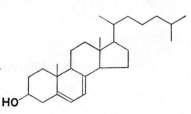

HO   HO

Cholesterol   Cholesta-5, 7-dien-3β-ol
                          (7-Dehydrocholesterol)

Azotobacter ozydans   Horváth, J. and A. Krámli, Nature, **160**, 639
                          (1947)

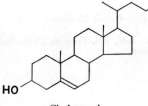   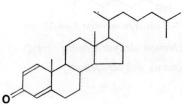

HO   O

Cholesterol   Cholesta-1, 4-dien-3-one

Arthrobacter simplex   Nagasawa, M., T. Hai, G. Tamura and K.
                          Arima, Abstr. paper (39th Meeting Agr. Chem.
                          Soc., Japan) p. 81, No. 415, 416 (1964)

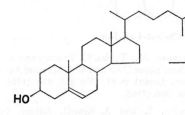   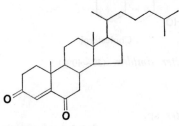

HO   O   O

Cholesterol   Cholest-4-ene-3, 6-dione

Mycobacterium cholesterolicum   Stadtman, T. C., A. Cherkes and C. B. Anfinsen,
                          J. Biol. Chem., **206**, 511 (1954)

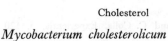

C. REDUCTION

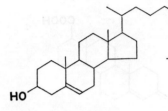
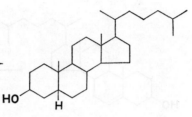

Cholesterol → 5β-Cholestan-3β-ol

Intestinal bacteria

Snog-kjaer, A., I. Prange and H. Dam, J. Gen. Microbiol., **14**, 256 (1956)

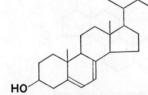
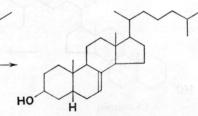

7-Dehydrocholesterol → 5β-Cholest-7-en-3β-ol

Fecal microorganisms

Coleman, D. L. and C. A. Baumann, Arch. Biochem. Biophys, **72**, 219 (1957)

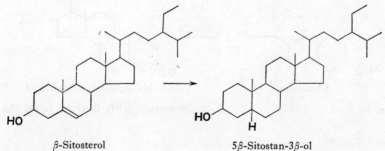

β-Sitosterol → 5β-Sitostan-3β-ol

Fecal microorganisms

Coleman, D. L. and C. A. Baumann, Arch. Biochem. Biophys., **72**, 219 (1957)

D. Side Chain Degradation

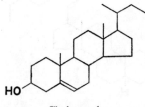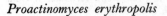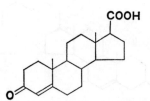

Cholesterol

3-Oxo-etiochol-4-enic acid
(Androst-4-en-3-one-17β-carboxylic acid)

Proactinomyces erythropolis　　　　　Turfitt, G. E., Biochem. J., **42**, 376 (1948)

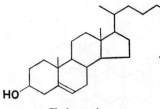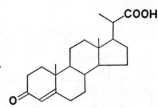

Cholesterol

3-Oxobisnorchol-4-enic acid

Nocardia sp.　　　　　Whitmarsh, J. M., Biochem. J., **90**, 23p. (1964)

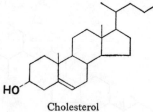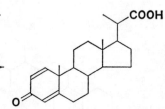

Cholesterol

3-Oxobisnorchola-1, 4-dienic acid

Nocardia sp.　　　　　Whitmarsh, J. M., Biochem. J., **90**, 23p. (1964)

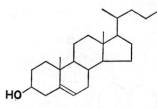

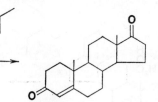

Cholesterol   Androst-4-ene-3, 17-dione

Nocardia sp. (low yield)

Arthrobacter simplex

Whitmarsh, J. M., Biochem. J., **90**, 23p. (1964)

Nagasawa, M., T. Hai, G. Tamura and K. Arima, Abstr. paper (39th and 40th Meeting Agr. Chem. Soc., Japan) p. 81, No. 415, 416 (1964) p. 63, No. 366 (1965)

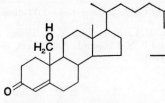

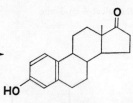

Cholesterol   Androsta-1, 4-diene-3, 17-dione

Nocardia sp. (low yield)

Arthrobacter simplex

Whitmarsh, J. M., Biochem. J., **90**, 23p. (1964)

Nagasawa, M., T. Hai, G. Tamura and K. Arima, Abstr. paper (39th and 40th Meeting Agr. Chem. Soc., Japan) p. 81, No. 415, 416 (1964) p. 63, No. 366 (1965)

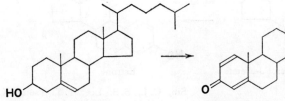

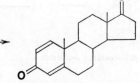

19-Hydroxycholest-4-en-3-one   Estrone

Nocardia restrictus ATCC 14887 (8%)

CSD-10 (30%)

Sih, C. J. and K. C. Wang, J. Am. Chem. Soc., **87**, 1387 (1965)

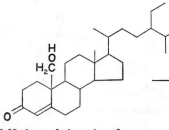

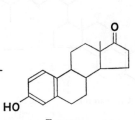

19-Hydroxy-β-sitost-4-en-3-one

Estrone

CSD-10 (10%)

Sih, C. J. and K. C. Wang, J. Am. Chem. Soc., **87**, 1387 (1965)

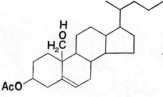

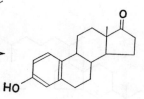

19-Hydroxycholesterol 3-acetate

Estrone

CSD-10 (72%)

Sih, C. J., S. S. Lee, Y. Y. Tsong, K. C. Wang and F. N. Chang, J. Am. Chem. Soc., **87**, 2765 (1965)

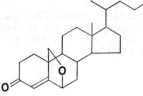

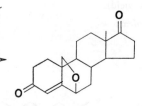

6β, 19-Oxidocholest-4-en-3-one

6β, 19-Oxido-androst-4-ene-3, 17-dione

CSD-10 (57%)

Sih, C. J., S. S. Lee, Y. Y. Tsong, K. C. Wang and F. N. Chang, J. Am. Chem. Soc., **87**, 2765 (1965)

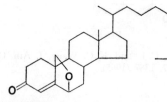

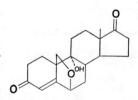

6β, 19-Oxidocholest-4-en-3-one

9α-Hydroxy-6β, 19-oxido-androst-4-ene-3, 17-dione

CSD-10

Sih, C. J., S. S. Lee, Y. Y. Tsong, K. C. Wang and F. N. Chang, J. Am. Chem. Soc., **87**, 2765 (1965)

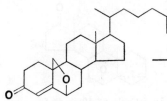 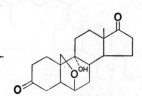

6β, 19-Oxidocholest-4-en-3-one

9α-Hydroxy-6β, 19-oxido-androstane-
3, 17-dione

CSD-10

Sih, C. J., S. S. Lee, Y. Y. Tsong, K. C. Wang
and F. N. Chang, J. Am. Chem. Soc., **87**, 2765
(1965)

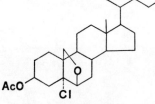 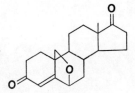

5α-Chloro-6β, 19-oxidocholestane
3-acetate

6β, 19-Oxido-androst-4-ene-3, 17-dione

CSD-10 (36%)

Sih, C. J., S. S. Lee, Y. Y. Tsong, K. C. Wang
and F. N. Chang, J. Am. Chem. Soc., **87**, 2765
(1965)

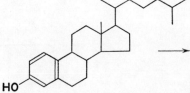 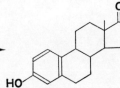

19-Norcholesta-1, 3, 5(10)-trien-3-ol

Estrone

Corynebacterium sp. (*Nocardia restrictus*) ATCC 14887 (8%)

Afonso, A., H. L. Herzog, C. Federbush and
W. Charney, Steroids, **7**, 429 (1966)

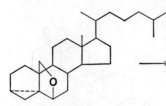 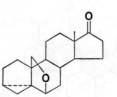

3α, 5-Cyclo-6β, 19-oxido-   3α, 5-Cyclo-6β, 19-oxido-5α-
      5α-cholestane   androstan-17-one

Corynebacterium equi B-58-1   Naito, A., M. Shirasaka, K. Tanabe, Abstr. paper
                                 (41st Meeting Agr. Chem. Soc., Japan) p. **139**,
                                 No. 756 (1966)

E. Cleavage of Steroid Skeleton

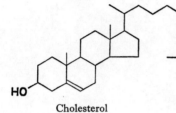 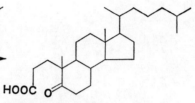

Cholesterol   Windaus keto acid
                               (5-Oxo-3, 5-seco-A-norcholestan-
                                        3-oic acid)

Proactinomyces erythropolis   Turfitt, G. E., Biochem. J., **42**, 376 (1948)

IV. MICROBIAL TRANSFORMATION OF SAPOGENINS

A. HYDROXYLATION

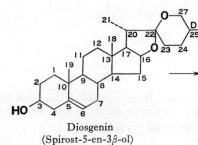

Diosgenin
(Spirost-5-en-3β-ol)

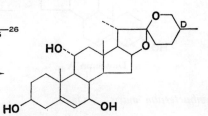

7β, 11α-Dihydroxydiosgenin

Helicostylum piriforme ATCC 8992 (10~ 15%)

Sato, Y. and S. Hayakawa, J. Org. Chem., **28**, 2742 (1963)

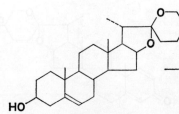

Diosgenin

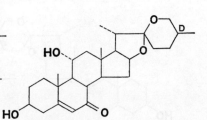

11α-Hydroxy-7-oxodiosgenin

Helicostylum piriforme ATCC 8992 (5~ 10%)

Sato, Y. and S. Hayakawa, J. Org. Chem., **28**, 2742 (1963)

B. DEHYDROGENATION

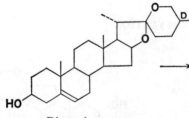
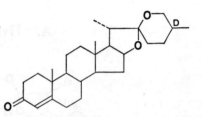

Diosgenin	Diosgenon
	(Spirost-4-en-3-one)

Brevibacterium maris A 126   Iizuka, H. and S. Iwado, Abstr. paper (40th Meeting Agr. Chem. Soc., Japan) p. 64 No. 367 (1965)

Corynebacterium simplex ATCC 6946   U. S. Pat. 3,134,718

Penicillium chrysogenum MF 2133 (3.6%)   Rothrock, J. W. and J. D. Garber, Arch. Biochem. Biophys, **57**, 151 (1955)

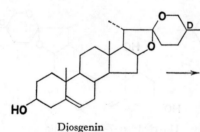
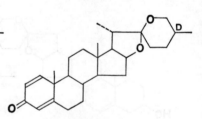

Diosgenin	Diosgedienon
	(Spirosta-1, 4-dien-3-one)

Brevibacterium maris A 126   Iizuka, H. and S. Iwado, Abstr. paper (40th Meeting Agr. Chem. Soc., Japan) p. 64 No. 367 (1965)

Corynebacterium simplex ATCC 6946   U. S. Pat. 3,134,718

C. Degradation of Spiroketal Ring

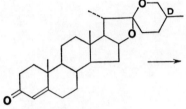
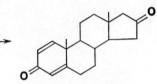

Diosgenon   Androsta-1, 4-diene-3, 16-dione

Fusarium solani No. 101 (65%)   Kondo, E. and T. Mitsugi, J. Am. Chem. Soc., 88, 4737 (1966)

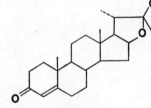
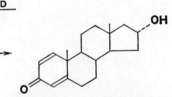

Diosgenon   16α-Hydroxyandrosta-1, 4-dien-3-one

Fusarium solani No. 101 (5%)   Kondo, E. and T. Mitsugi, J. Am. Chem. Soc., 88, 4737 (1966)

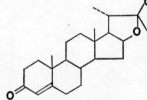
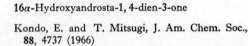
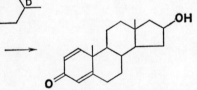

Diosgenon   16β-Hydroxyandrosta-1, 4-dien-3-one

Fusarium solani No. 101 (5%)   Kondo, E. and T. Mitsugi, J. Am. Chem. Soc., 88, 4737 (1966)

V. MICROBIAL TRANSFORMATION OF CARDENOLIDES AND BUFADIENOLIDES

Structures of typical cardenolides and bufadienolide

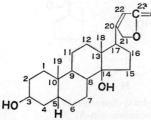

Digitoxigenin
(3β, 14-Dihydroxy-5β-card-
20(22)-enolide)

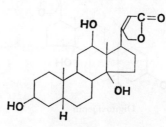

Digoxigenin
(3β, 12β, 14-Trihydroxy-5β-card-
20(22)-enolide)

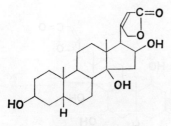

Gitoxigenin
(3β, 14, 16β-Trihydroxy-5β-card-
20(22)-enolide)

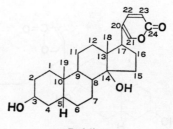

Bufalin
(3β, 14-Dihydroxy-5β-bufa-
20, 22-dienolide)

A. Hydroxylation

(a) 1-Hydroxylation

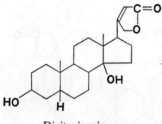

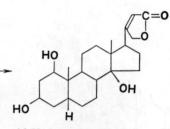

Digitoxigenin	1β-Hydroxydigitoxigenin (Acovenosigenin-A)

Absidia orchidis

Nozaki Y. and T. Okumura, Agr. Chem. Soc. (Japan), **25**, 515 (1961)

Rhizopus nigricans ATCC 6227b

Nozaki, Y., E. Masuo and D. Satoh, Agr. Chem. Soc. (Japan), **26**, 399 (1962)

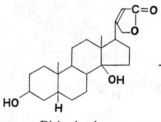

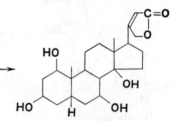

Digitoxigenin	1β, 7β-Dihydroxydigitoxigenin

Absidia orchidis

Ishii, H., Y. Nozaki, T. Okumura and D. Satoh, J. Pharm. Soc. (Japan), **81**, 1051 (1961)

(b) 5-Hydroxylation

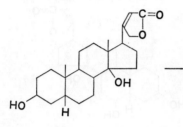 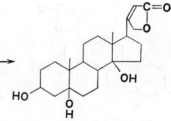

Digitoxigenin

5β-Hydroxydigitoxigenin
(Periplogenin)

Absidia orchidis

Ishii, H., Y. Nozaki, T. Okumura and D. Satoh, J. Pharm. Soc. (Japan), **81**, 1051 (1961)

Mucor parasiticus ATCC 6476

Ishii, H., J. Pharm. Soc. (Japan), **81**, 153, (1961)

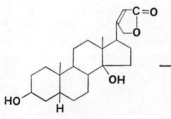 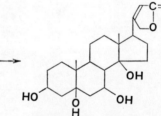

Digitoxigenin

5β, 7β-Dihydroxydigitoxigenin

Absidia orchidis

Ishii, H., Y. Nozaki, T. Okumura and D. Satoh, J. Pharm. Soc. (Japan), **81**, 1051 (1961)

(c) 6-Hydroxylation

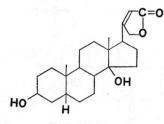

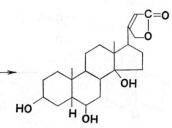

Digitoxigenin

6β-Hydroxydigitoxigenin

Trichothecium roseum ATCC 8685

Titus, E., A. W. Murray and H. E. Spiegel, J. Biol. Chem., **235**, 3399 (1960)

(d) 7-Hydroxylation

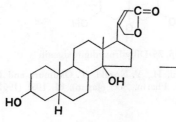

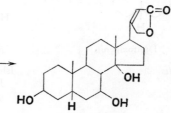

Digitoxigenin   7β-Hydroxydigitoxigenin

Absidia orchidis (4%)

Nozaki, Y., E. Masuo and D. Satoh, Agr. Biol. Chem. (Japan), **26**, 399 (1962)

Aspergillus oryzae

Juhasz, G. and Ch. Tamm, Helv. Chim. Acta., **44**, 1063 (1961)

Mucor parasiticus ATCC 6476 (2%)

Nozaki, Y., E. Masuo and D. Satoh, Agr. Biol. Chem. (Japan), **26**, 399 (1962)

Rhizopus arrhizus ATCC 11145

Ishii, H., Y. Nozaki, T. Okumura and D. Satoh, J. Pharm. Soc. (Japan), **81**, 805 (1961)

Rhizopus nigricans ATCC 6227b (59%)

Nozaki, Y., E. Masuo and D. Satoh, Agr. Biol. Chem. (Japan), **26**, 399 (1962)

Trichothecium roseum
(Cephalothecium roseum)

Juhasz, G. and Ch. Tamm, Helv. Chim. Acta., **44**, 1063 (1961)

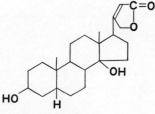

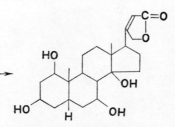

Digitoxigenin   1β,7β-Dihydroxydigitoxigenin

Absidia orchidis

Ishii, H., Y. Nozaki, T. Okumura and D. Satoh, J. Pharm. Soc. (Japan), **81**, 1051 (1961)

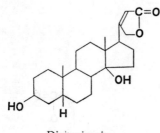 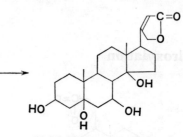

Digitoxigenin   5β,7β-Dihydroxydigitoxigenin

Absidia orchidis   Ishii, H., Y. Nozaki, T. Okumura and D. Satoh,
                                 J. Pharm. Soc. (Japan), **81**, 1051 (1961)

(e) 11-Hydroxylation

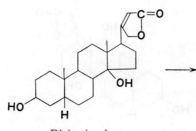 →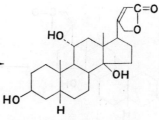

Digitoxigenin   11α-Hydroxydigitoxigenin
(Sarmentogenin)

Aspergillus ochraceus

Nozaki, Y. and K. Akagi, 39th Meeting Agr. Chem. Soc. (Japan), Abstr. paper p. 100 No. 463 (1964)

Trichothecium roseum ATCC 8685

Titus, E., A. W. Murray and H. E. Spiegel, J. Biol. Chem., **235**, 3399 (1960)

(f) 12-Hydroxylation

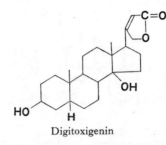 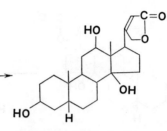

Digitoxigenin   12β-Hydroxydigitoxigenin
(Digoxigenin)

Calonectria decora   Nozaki, Y., E. Masuo, H. Ishii, T. Okumura and D. Satoh, Abstr. Paper (Symposium on the Chem. of Digitalis Cardiac Glycosides, Tokyo) p. 114 (1960)

Fusarium lini   Gubler, A. and Ch. Tamm, Helv. Chim. Acta., **41**, 297 (1958)

Gibberella fujikuroii   Nawa, H., M. Uchibayashi, T. Kamiya, T. Yamano, H. Arai and M. Abe, Nature, **184**, 469 (1959)

Gibberella saubinetti (70%)   Okada, M., A. Yamada and M. Ishidate, Chem. Pharm. Bull. (Japan), **8**, 530 (1960)

Helicostylum piriforme   Nawa, H., M. Uchibayashi, T. Kamiya, T. Yamano, H. Arai and M. Abe Nature, **184**, 469 (1959)

Nigrospora sphaerica   Nozaki, Y., E. Masuo, H. Ishii, T. Okumura and D. Satoh, Abstr. Paper (Symposium on the Chem. of Digitalis Cardiac Glycosides, Tokyo) p. 114 (1960)

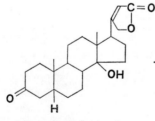 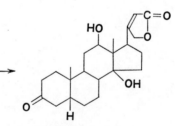

3-Dehydrodigitoxigenin   3-Dehydrodigoxigenin

Fusarium lini   Gubler, A. and Ch. Tamm, Helv. Chim. Acta., **41**, 297 (1958)

Gibberella saubinetti (53%)   Okada, M., A. Yamada and M. Ishidate, Chem. Pharm. Bull. (Japan), **8**, 530 (1960)

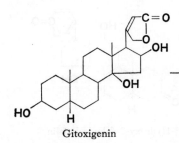
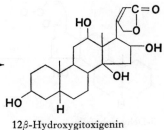

Gitoxigenin

12β-Hydroxygitoxigenin
(Diginatigenin)

Fusarium lini (0.5%)

Tamm, Ch. and A. Gubler, Helv. Chim. Acta.,
41, 1762 (1958)

Gibberella saubinetti (6%)

Okada, M., A. Yamada and M. Ishidate, Chem.
Pharm. Bull. (Japan), **8**, 530 (1960)

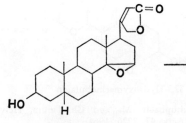
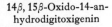
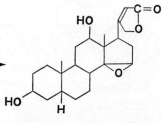

14β, 15β-Oxido-14-an-
hydrodigitoxigenin

12β-Hydroxy-14β, 15β-oxido-
14-anhydrodigitoxigenin

Fusarium lini

Schüpbach, M. and Ch. Tamm, Helv. Chim.
Acta., **47**, 2217 (1964)

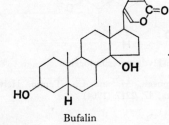
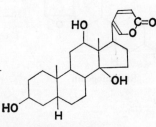

Bufalin

12β-Hydroxybufalin

Fusarium lini

Tamm, Ch. and A. Gubler, Helv. Chim. Acta,
42, 473 (1959)

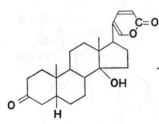 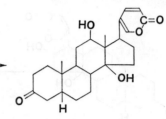

3-Dehydrobufalin   12β-Hydroxy-3-dehydrobufalin

Fusarium lini

Tamm, Ch. and A. Gubler, Helv. Chim. Acta, **42**, 473 (1959)

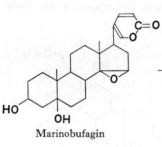 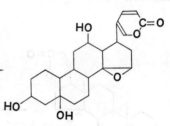

Marinobufagin   12β-Hydroxymarinobufagin

Fusarium lini

Schüpbach, M. and Ch. Tamm, Helv. Chim. Acta, **47**, 2226 (1964)

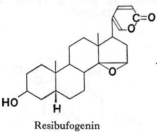 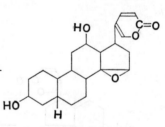

Resibufogenin   12β-Hydroxyresibufogenin

Fusarium lini

Schüpbach, M. and Ch. Tamm, Helv. Chim. Acta, **47**, 2217 (1964)

(g) 16-Hydroxylation

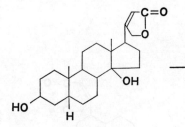

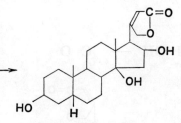

Digitoxigenin

Gitoxigenin

Cunninghamella blakesleeana
Helicostylum piriforme

Nawa, H., M. Uchibayashi, T. Kamiya, T. Yamano, H. Arai and M. Abe, Nature, **184**, 469
(1959)

B. Dehydrogenation

(a) $>$CH–OH $\longrightarrow$ $>$C=O

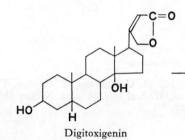

Digitoxigenin

Aspergillus oryzae
Calonectria decora (16%)
Nigrospora sphaerica (12%)

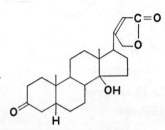

3-Dehydrodigitoxigenin

Nozaki, Y., E. Masuo, H. Ishii, T. Okumura and D. Satoh, Abstr. Paper (Symposium on the Chem. of Digitalis Cardiac Glycosides, Tokyo) p. 114 (1960)

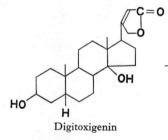

Digitoxigenin

Rhizopus arrhizus

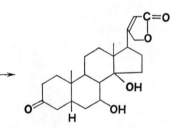

7β-Hydroxy-3-dehydrodigitoxigenin

Juhasz, G. and Ch. Tamm, Helv. Chim. Acta, **44**, 1063 (1961)

(b) $-CH_2-CH\bigg\langle \longrightarrow -CH=C\bigg\langle$

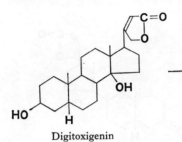

Digitoxigenin

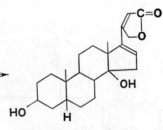

16-Dehydrodigitoxigenin
(Δ^{16}-Anhydrogitoxigenin)

Trichothecium roseum ATCC 8685

Titus, E., A. W. Murray and H. E. Spiegel, J. Biol. Chem., **235**, 3399 (1960)

C. REDUCTION

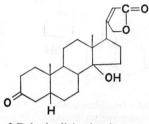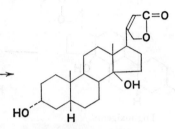

3-Dehydrodigitoxigenin → 3-Epidigitoxigenin

Fusarium lini

Gubler, A. and Ch. Tamm, Helv. Chim. Acta, **41**, 297 (1958)

Gibberella saubinetti (3.5%)

Okada, M., A. Yamada and M. Ishidate, Chem. Pharm. Bull. (Japan), **8**, 530 (1960)

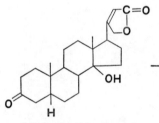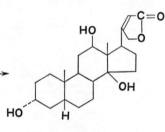

3-Dehydrodigitoxigenin → 3-Epidigoxigenin

Fusarium lini

Gubler, A. and Ch. Tamm, Helv. Chim. Acta, **41**, 297 (1958)

Gibberella saubinetti

Okada, M., A. Yamada and M. Ishidate, Chem. Pharm. Bull. (Japan), **8**, 530 (1960)

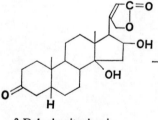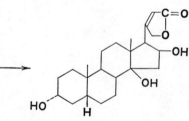

3-Dehydrogitoxigenin → 3-Epigitoxigenin

Fusarium lini

Tamm, Ch. and A. Gubler, Helv. Chim. Acta, **41**, 1762 (1958)

— 244 —

VI. MICROBIAL TRANSFORMATION OF STEROIDAL ALKALOIDS

Structures of typical steroidal alkaloids

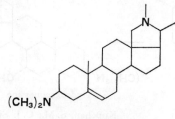

(CH₃)₂N

Conessine

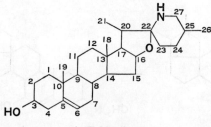

Solasodine

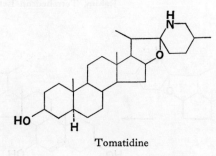

Tomatidine

A. HYDROXYLATION

(a) 7-Hydroxylation

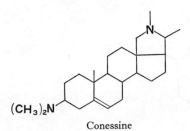
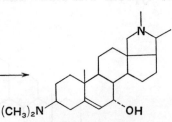

Conessine

7α-Hydroxyconessine

Aspergillus ochraceus

Kupchan, S. M., C. J. Sih, S. Kubota and A. M.
Rahim, Tetrahedron Letter No. 26, 1767 (1963)

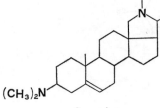
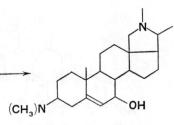

Conessine

7β-Hydroxyconessine

Aspergillus ochraceus

Kupchan, S. M., C. J. Sih, S. Kubota and A. M.
Rahim, Tetrahedron Letter No. 26, 1767 (1963)

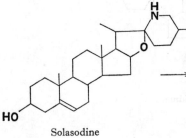
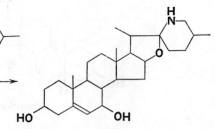

Solasodine

7β-Hydroxysolasodine

Helicostylum piriforme   ATCC 8992 (1%)

Sato, Y. and S. Hayakawa, J. Org. Chem., **28**,
2739 (1963)

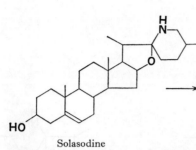
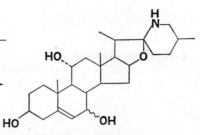

Solasodine   7ξ, 11α-Dihydroxysolasodine

Helicostylum piriforme ATCC 8992
(0.5%)

Sato, Y. and S. Hayakawa, J. Org. Chem., **28,**
2739 (1963)

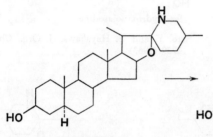
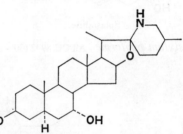

Tomatidine   7α-Hydroxytomatidine

Helicostylum piriforme ATCC 8992 (5%)

Sato, Y. and S. Hayakawa, J. Org. Chem., **29,**
198 (1964)

(b) 9-Hydroxylation

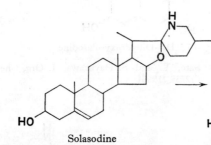 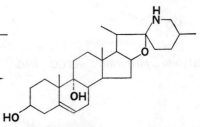

Solasodine   9α-Hydroxysolasodine

Helicostylum piriforme ATCC 8992 (30~ 35%) Sato, Y. and S. Hayakawa, J. Org. Chem., **28**, 2739 (1963)

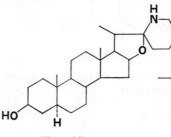 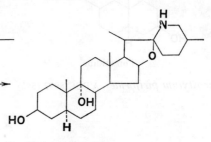

Tomatidine   9α-Hydroxytomatidine

Helicostylum piriforme ATCC 8992 (0.5%) Sato, Y. and S. Hayakawa, J. Org. Chem., **29**, 198 (1964)

(c) 11-Hydroxylation

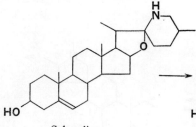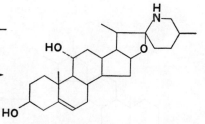

Solasodine

11α-Hydroxysolasodine

Helicostylum piriforme ATCC 8992 (1%)

Sato, Y. and S. Hayakawa, J. Org. Chem., **28**, 2739 (1963)

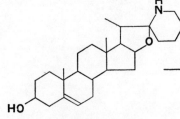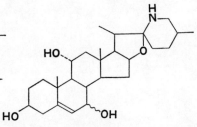

Solasodine

7ξ, 11α-Dihydroxysolasodine

Helicostylum   piriforme   ATCC   8992 (0.5%)

Sato, Y. and S. Hayakawa, J. Org. Chem., **28**, 2739 (1963)

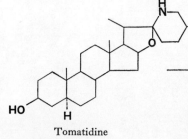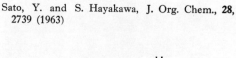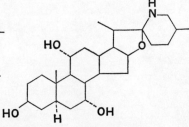

Tomatidine

7α, 11α-Dihydroxytomatidine

Helicostylum piriforme ATCC 8992 (25~ 30%)

Sato, Y. and S. Hayakawa, J. Org. Chem., **29**, 198 (1964)

B. DEHYDROGENATION

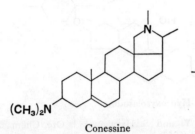

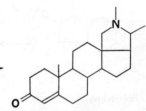

Conessine

4-Dehydroconenin-3-one

Gloeosporium cyclaminis
Hypomyces haematococcus

De Flines, J., A. F. Marx, W. F. van der Waad and D. van der Sijde, Tetrahedron Letter No. 26, 1257 (1962)

VII. MICROBIAL TRANSFORMATION OF ERGOT ALKALOIDS

A. HYDROXYLATION

(a) 8-Hydroxylation

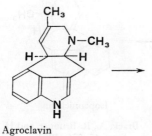 →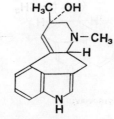

Agroclavin   Setoclavin

Psilocybe semperviva   Brack, A., R. Brunner and H. Kobel, Helv.
                              Chim. Acta, **45**, 276 (1962)

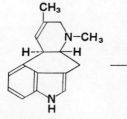 →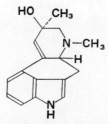

Agroclavin   Isosetoclavin

Psilocybe semperviva   Brack, A., R. Brunner and H. Kobel, Helv.
                              Chim. Acta, **45**, 276 (1962)

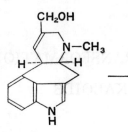

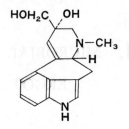

Elymoclavin   Penniclavin

Psilocybe semperviva   Brack, A., R. Brunner and H. Kobel, Helv. Chim.
                                       Acta., **45**, 276 (1962)

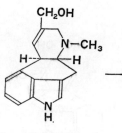

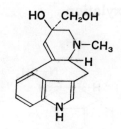

Elymoclavin   Isopenniclavin

Psilocybe semperviva   Brack, A., R. Brunner and H. Kobel, Helv. Chim.
                                       Acta, **45**, 276 (1962)

VIII. MICROBIAL TRANSFORMATION OF INDOL ALKALOIDS

Structures of typical indol alkaloids

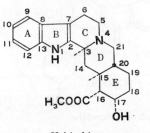

Yohimbine

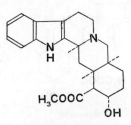

α-Yohimbine

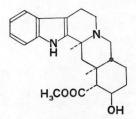

β-Yohimbine

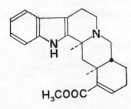

Apoyohimbine

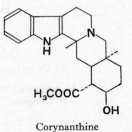

Corynanthine

A. HYDROXYLATION

(a) 10-Hydroxylation

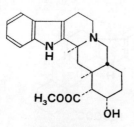

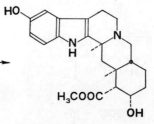

Yohimbine

10-Hydroxyyohimbine

Cunninghamella bainieri ATCC 924
Cunninghamella blakesleeana ATCC 8688a
Cunninghamella echinulata NRRL A-11498
Streptomyces platensis NRRL 2364
Streptomyces rimosus NRRL 2234

Hartman, R. E., E. F. Krause, W. W. Andres and E. L. Patterson, Appl. Microbiol., **12**, 138 (1964)

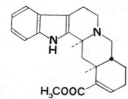

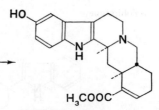

Apoyohimbine

10-Hydroxyapoyohimbine

Cunninghamella blakesleeana

Godtfredsen, W. O., G. Korsby, H. Lorck and S. Vangedal, Experientia, **14**, 88 (1958)

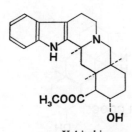

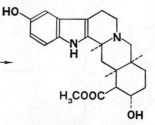

α-Yohimbine · 10-Hydroxy-α-yohimbine

Cunninghamella bainieri ATCC 924
Cunninghamella blakesleeana ATCC 8688a
Cunninghamella echinulata NRRL A-11498
Streptomyces platensis NRRL 2364

Hartman, R. E., E. F. Krause, W. W. Andres and E. L. Patterson, Appl. Microbiol., **12**, 138 (1964)

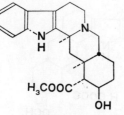

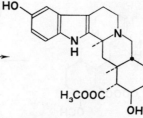

β-Yohimbine · 10-Hydroxy-β-yohimbine

Cunninghamella bainieri ATCC 924
Cunninghamella echinulata NRRL A-11498
Streptomyces platensis NRRL 2364
Streptomyces rimosus NRRL 2234

Hartman, R. E., E. F. Krause, W. W. Andres and E. L. Patterson, Appl. Microbiol., **12**, 138 (1964)

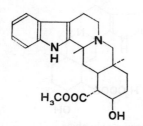
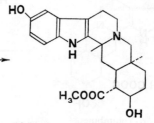

Corynanthine   10-Hydroxycorynanthine

Cunninghamella bainieri ATCC 924   Hartman, R. E., E. F. Krause, W. W. Andres and
Cunninghamella echinulata NRRL A-   E. L. Patterson, Appl. Microbiol., **12**, 138 (1964)
11498

Streptomyces platensis NRRL 2364

Streptomyces rimosus NRRL 2234

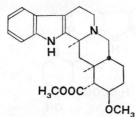
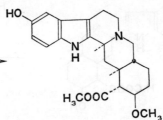

β-Yohimbine methylether   10-Hydroxy-β-yohimbine methylether

Cunninghamella blakesleeana   Godtfredsen, W. O., G. Korsby, H. Lorck and
                                       S. Vangedal, Experientia, **14**, 88 (1958)

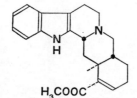
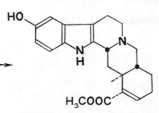

3-Epiapoyohimbine   10-Hydroxy-3-epiapoyohimbine

Cunninghamella blakesleeana   Godtfredsen, W. O., G. Korsby, H. Lorck and
                                       S. Vangedal, Experientia, **14**, 88 (1958)

(b) 11-Hydroxylation

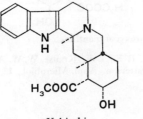 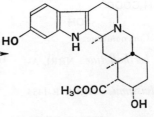

Yohimbine   11-Hydroxyyohimbine

Cunninghamella bainieri strain Campbell X-48

Cunninghamella bertholletiae NRRL A-11497

Cunninghamella echinulata

Cunninghamella elegans NRRL A-11499

Hartman, R. E., E. F. Krause, W. W. Andres and E. L. Patterson, Appl. Microbiol., **12**, 138 (1964)

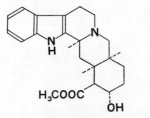 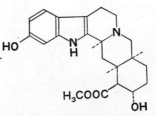

α-Yohimbine   11-Hydroxy-α-yohimbine

Cunninghamella bainieri strain Campbell X-48

Cunninghamella bertholletiae NRRL A-11497

Cunninghamella echinulata

Cunninghamella elegans NRRL A-11499

Streptomyces fulvissimus NRRL B-1453

Hartman, R. E., E. F. Krause, W. W. Andres and E. L. Patterson, Appl. Microbiol., **12**, 138 (1964)

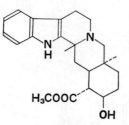 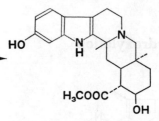

Corynanthine   11-Hydroxycorynanthine

Cunninghamella bertholletiae NRRL A-
11497

Streptomyces fulvissimus NRRL B-1453

Hartman, R. E., E. F. Krause, W. W. Andres and
E. L. Patterson, Appl. Microbiol., **12**, 138 (1964)

(c) 18-Hydroxylation

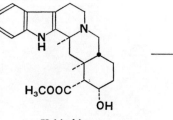

Yohimbine

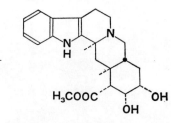

18α-Hydroxyyohimbine

Calonectria decora

Hartman, R. E., E. F. Krause, W. W. Andres and
E. L. Patterson, Appl. Microbiol., **12**, 138 (1964)

Streptomyces aureofaciens ATCC 11834
Streptomyces rimosus NRRL 2234

Pan, S. C. and F. L. Weisenborn, J. Am. Chem.
Soc., **80**, 4749 (1958)

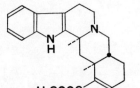

Apoyohimbine

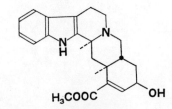

18-Hydroxyapoyohimbine

Cunninghamella blakesleeana

Godtfredsen, W. O., G. Korsby, H. Lorck and
S. Vangedal, Experientia, **14**, 88 (1958)

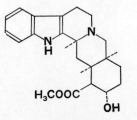

α-Yohimbine

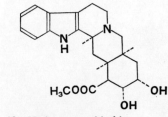

18α-Hydroxy-α-yohimbine

Streptomyces aureofaciens ATCC 11834
Streptomyces rimosus NRRL 2234

Pan, S. C. and F. L. Weisenborn, J. Am. Chem.
Soc., **80**, 4749, (1958)

IX. MICROBIAL TRANSFORMATION OF MORPHINE ALKALOIDS

Structures of typical morphine alkaloids

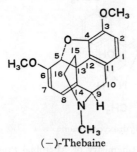

(−)-Thebaine

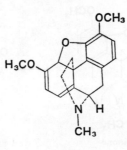

(+)-Thebaine

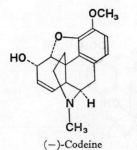

(−)-Codeine

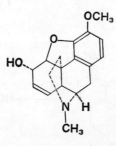

(+)-Codeine

A. HYDROXYLATION

(a) 14-Hydroxylation

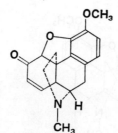

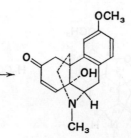

(+)-Codeinone

(+)-14-Hydroxycodeinone

Trametes sanguinea (10.9%)

Tsuda, K., Chemistry of Microbial Products (6th Symposium of the Inst. Appl. Microbiol. Univ. of Tokyo) p. 167 (1964)

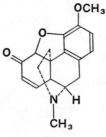

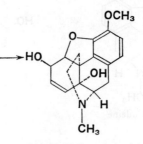

(+)-Codeinone

(+)-14-Hydroxyisocodeine

Trametes sanguinea (3.2%)

Tsuda, K., Chemistry of Microbial Products (6th Symposium of the Inst. Appl. Microbiol. Univ. of Tokyo) p. 167 (1964)

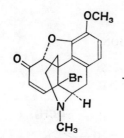

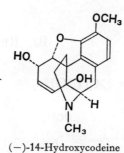

(−)-14-Bromocodeinone

(−)-14-Hydroxycodeine

Trametes sanguinea (16%)

Yamada, M., K. Iizuka, S. Okuda, T. Asai and K. Tsuda, Chem. Pharm. Bull. (Japan), **11**, 206 (1963)

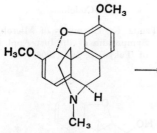

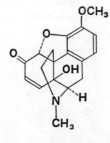

(−)-Thebaine

(−)-14-Hydroxycodeinone

Trametes sanguinea (Polystictus sanguineus) (40%)

Iizuka, K., M. Yamada, J. Suzuki, I. Seki, K. Aida, S. Okuda, T. Asai and K. Tsuda, Chem. Pharm. Bull. (Japan), **10**, 67 (1962)

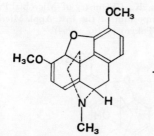

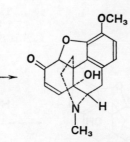

(+)-Thebaine

(+)-14-Hydroxycodeinone

Trametes sanguinea (40%)

Tsuda, K., Chemistry of Microbial Products (6th Symposium of the Inst. Appl. Microbiol. Univ. of Tokyo) p. 167 (1964)

B. REDUCTION

(a) $\text{>C=O} \longrightarrow \text{>CH—OH}$

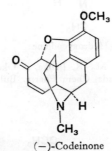

(−)-Codeinone

Trametes sanguinea (24.5%)

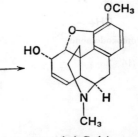

(−)-Codeine

Tsuda, K., Chemistry of Microbial Products (6th Symposium of the Inst. Appl. Microbiol. Univ. of Tokyo) p. 167 (1964)

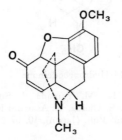

(+)-Codeinone

Trametes sanguinea

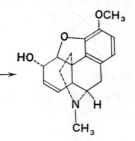

(+)-Codeine

Tsuda, K., Chemistry of Microbial Products (6th Symposium of the Inst. Appl. Microbiol. Univ. of Tokyo) p. 167 (1964)

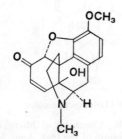

(−)-14-Hydroxycodeinone

Trametes sanguinea (*Polystictus sanguineus*) (55%)

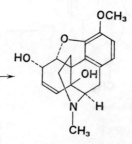

(−)-14-Hydroxycodeine

Iizuka, K., M. Yamada, J. Suzuki, I. Seki, K. Aida, S. Okuda, T. Asai and K. Tsuda, Chem. Pharm. Bull. (Japan), **10**, 67 (1962)

B. REDUCTION

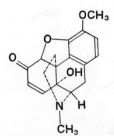
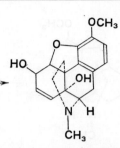

(+)-14-Hydroxycodeinone

Trametes sanguinea

(+)-14-Hydroxyisocodeine

Tsuda, K., Chemistry of Microbial Products (6th Symposium of the Inst. Appl. Microbiol. Univ. of Tokyo) p. 167 (1964)

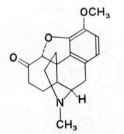

(−)-Dihydrocodeinone

Trametes sanguinea (31%)

(−)-Dihydrocodeine

Yamada, M., Chem. Pharm. Bull. (Japan), **11**, 356 (1963)

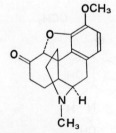
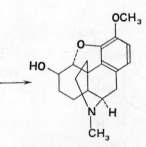

(−)-Dihydrocodeinone

Trametes sanguinea (10%)

(−)-Dihydroisocodeine

Tsuda, K., Chemistry of Microbial Products (6th Symposium of the Inst. Appl. Microbiol. Univ. of Tokyo) p. 167 (1964)

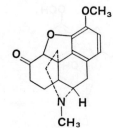 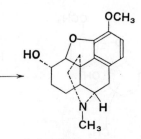

(+)-Dihydrocodeinone

Trametes sanguinea (2.7%)

(+)-Dihydrocodeine

Tsuda, K., Chemistry of Microbial Products (6th Symposium of the Inst. Appl. Microbiol. Univ. of Tokyo) p. 167 (1964)

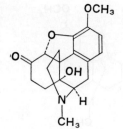 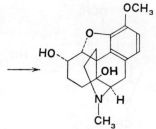

(−)-14-Hydroxydihydrocodeinone

Trametes sanguinea (38.4%)

(−)-14-Hydroxydihydrocodeine

Tsuda, K., Chemistry of Microbial Products (6th Symposium of the Inst. Appl. Microbiol. Univ. of Tokyo) p. 167 (1964)

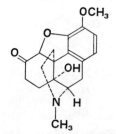 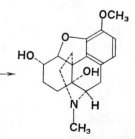

(+)-14-Hydroxydihydrocodeinone

Trametes sanguinea

(+)-14-Hydroxydihydrocodeine

Tsuda, K., Chemistry of Microbial Products (6th Symposium of the Inst. Appl. Microbiol. Univ. of Tokyo) p. 167 (1964)

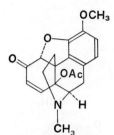

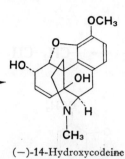

(−)-14-Acetoxycodeinone

(−)-14-Hydroxycodeine

Trametes sanguinea (70%)

Yamada, M., K. Iizuka, S. Okuda, T. Asai and K. Tsuda, Chem. Pharm. Bull. (Japan), **11**, 206 (1963)

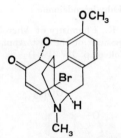

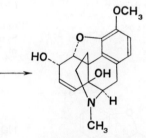

(−)-14-Bromocodeinone

(−)-14-Hydroxycodeine

Trametes sanguinea (16%)

Yamada, M., K. Iizuka, S. Okuda, T. Asai and K. Tsuda, Chem. Pharm. Bull. (Japan), **11**, 206 (1963)

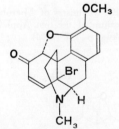

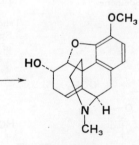

(−)-14-Bromocodeinone

(−)-Neopine

Trametes sanguinea (4.3%)

Yamada, M., K. Iizuka, S. Okuda, T. Asai and K. Tsuda, Chem. Pharm. Bull. (Japan), **11**, 206 (1963)

(b) $-CH=CH- \longrightarrow -CH_2-CH_2-$

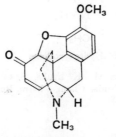

(+)-Codeinone

Trametes sanguinea (4.9%)

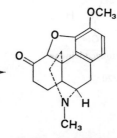

(+)-Dihydrocodeinone

Tsuda, K., Chemistry of Microbial Products (6th Symposium of the Inst. Appl. Microbiol. Univ. of Tokyo) p. 167 (1964)

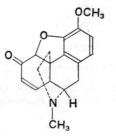

(+)-Codeinone

Trametes sanguinea

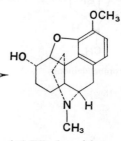

(+)-Dihydrocodeine

Tsuda, K., Chemistry of Microbial Products (6th Symposium of the Inst. Appl. Microbiol. Univ. of Tokyo) p. 167 (1964)

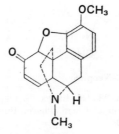

(+)-Codeinone

Trametes sanguinea

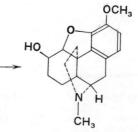

(+)-Dihydroisocodeine

Tsuda, K., Chemistry of Microbial Products (6th Symposium of the Inst. Appl. Microbiol. Univ. of Tokyo) p. 167 (1964)

C. HYDROLYSIS

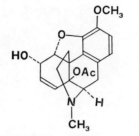

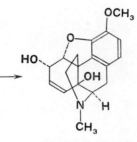

(−)-14-Acetoxycodeine

Trametes sanguinea (70%)

(−)-14-Hydroxycodeine

Yamada, M., K. Iizuka, S. Okuda, T. Asai and
K. Tsuda, Chem. Pharm. Bull. (Japan), **11**, 206
(1963)

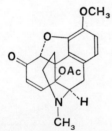

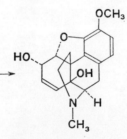

(−)-14-Acetoxycodeinone

Trametes sanguinea (70%)

(−)-14-Hydroxycodeine

Yamada, M., K. Iizuka, S. Okuda, T. Asai and
K. Tsuda, Chem. Pharm. Bull. (Japan), **11**, 206
(1963)

X. MICROBIAL TRANSFORMATION

OF NICOTINE

Nicotine

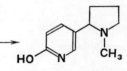

1-Methyl-2(6-hydroxy-3-pyridyl)-
pyrrolidine

Unidentified soil bacterium

Hochstein, L. I. and S. C. Rittenberg, J. Biol.
Chem., **235**, 795 (1960)

Nicotine

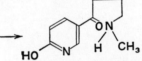

6-Hydroxypseudooxonicotine

Unidentified soil bacterium

Hochstein L. I. and S. C. Rittenberg, J. Biol.
Chem., **235**, 795 (1960)

Nicotine

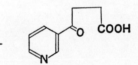

3-Succinoylpyridine

Achromobacter denitrificans
Bacillus megaterium var. *nicotinovorus*
Bacterium flavescens
Bacterium mutabile var. *acidoformans*
Bacterium qualis var. *amylophilum*
Pseudomonas nicotinophaga
Pseudomonas cyclosites
Pseudomonas nicotiana
Xanthomonas carotae var. *nicotinovora*

Tabuchi, T., J. Agr. Chem. Soc. (Japan), **28**, 806
(1954)

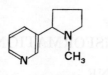 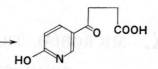

Nicotine   3-Succinoylpyridone-6

Achromobacter denitrificans

Bacillus megaterium var. *nicotinovorus*

Bacterium flavescens

Bacterium mutabile var. *acidoformans*

Bacterium qualis var. *amylophilum*

Pseudomonas cyclosites

Pseudomonas nicotiana

Pseudomonas nicotinophaga

Xanthomonas carotae var. *nicotinovora*

Tabuchi, T., J. Agr. Chem. Soc. (Japan), **28**, 806 (1954)

Author Index

Microorganism Index

Substance Index

DATE DUE

~~JAN 2 7 1978~~		
~~FEB - 4 1978~~ FEB - 4 1978	MAY - 1 1978	
MAR 25 1979		
APR 22 1979	~~APR~~ 9 1979	
261-2500		Printed in USA